SUSAN MYERS

# BIRDS OF
# BORNEO

## A PHOTOGRAPHIC GUIDE

H E L M

LONDON · OXFORD · NEW YORK · NEW DELHI · SYDNEY

HELM
Bloomsbury Publishing Plc
50 Bedford Square, London, WC1B 3DP, UK
29 Earlsfort Terrace, Dublin 2, Ireland

BLOOMSBURY, HELM and the Helm logo are trademarks
of Bloomsbury Publishing Plc

This edition published 2023

A catalogue record for this book is available from the British Library
Library of Congress Cataloguing-in-Publication data has been applied for

ISBN: PB: 978-1-4729-8690-0; ePub: 978-1-4729-8689-4;
ePDF: 978-1-4729-9387-8

2 4 6 8 10 9 7 5 3 1

Design by Rod Teasdale
Printed and bound in India by Replika Press Pvt. Ltd.

To find out more about our authors and books visit www.bloomsbury.com
and sign up for our newsletters

# CONTENTS

# INTRODUCTION

## Geography

Borneo is the world's third largest island, after Greenland and New Guinea. It lies east of Sumatra and the Malay Peninsula, north of Java, south of the Philippines and west of Sulawesi, in the centre of maritime South-East Asia. The island straddles the equator in a key position on the edge of the South-East Asian region adjacent to the interchange zone ('Wallacea') with the Australasian region. It is divided between three countries – Brunei, Malaysia and Indonesia. The provinces of Kalimantan (Indonesia) occupy the major portion of the island's surface area, c.73%, while the states of Sabah and Sarawak (Malaysia) cover approximately 26% of its area. Brunei occupies just 1% of the area of Borneo.

Geologically, the Thai-Malay Peninsula and much of the Indo-Malayan archipelago lie on the Sunda Shelf, an extension of the continental shelf of South-East Asia. The term 'Sundaland', or Sundaic Region, refers to the Thai-Malay Peninsula, as well as Sumatra, Java, Bali, Borneo, Palawan and many smaller islands on the Sunda Shelf.

While much of Borneo is low-lying – well over half of its area lies below 150m – a central mountain range extends from Sabah in the north in a south-westerly direction,

*Mount Kinabalu*

Relief map of Borneo showing major mountains and highland areas, and other geographical features.

1. Gunung Kinabalu
2. Trus Madi Range
3. Crocker Range
4. Labuk Highlands (Gunung Meliau)
5. Maliau Basin
6. Danum Valley
7. Bagabak Hills
8. Tawau Hills
9. Witti Range (Gunung Napotong)
10. Meligan Range
11. Kelabit Highlands
12. Kayan Mentarang
13. Gunung Mulu
14. Dulit Range
15. Usun Apau Plateau
16. Iran Range (Batu Tibang)
17. Hose Mountains
18. Gunung Menyapa
19. Bukit Liang Pran (Batu Tibang)
20. Gunung Lawit
21. Kapuas Hulu Range
22. Klingkang Range
23. Gunung Niut and Penrissen
24. Pueh Range
25. Barito Ulu
26. Danau Sentarum
27. Muller Range (Gunung Liang Kubung)
28. Gunung Saran
29. Gunung Palung
30. Schwaner Range
31. Mahakam Lakes
32. Meratus Mountains

along the border between Sarawak and Kalimantan to roughly the centre of Borneo. The central ranges and outliers are generally relatively low. At 4,095m, Mount (Gunung) Kinabalu is Borneo's highest peak and is also the highest mountain between West Papua and the Himalayas; few other peaks on Borneo exceed 2,000m, with all of the highest mountains in Sabah. These also host the largest number of endemic species. Throughout Borneo a vast network of rivers rises from the interior mountains and fans down to the coast. These extensive waterways are important not only for wildlife, but for transport, trade and tourism. Most of the human settlements on Borneo are concentrated on the coastline and along the rivers.

## Climate

Borneo lies between 07°N and 04°S, with the equator passing roughly through the centre of the island. The island has a tropical moist climate with heavy rainfall in all months and high humidity. Temperatures are stable, rarely fluctuating more than 10°C, with means of 25°–35°C in the lowlands throughout the year. The island is influenced by two monsoonal systems. The north-west monsoon, from November to April, is wetter than the so-called dry south-east monsoon from May to October. Monthly rainfall generally exceeds 200mm but peaks in November, with a second peak in April. June–August are the driest months, but even then, rainfall is never less than 100mm a month.

## Biodiversity

The flora of Borneo is amongst the most diverse in the world with 10,000–15,000 species of flowering plants and 3,000 species of trees, many of them endemic. There is a correspondingly rich and diverse fauna with more than 280 species of mammals, 440 species of freshwater fishes, almost 300 species of reptiles and more than 650 species of birds. Moreover, a high percentage of each of these groups is endemic, in other words, they are unique to Borneo. There are at least 44 endemic mammals, 100 amphibians, almost 100 reptiles and 52 birds.

Much of Borneo's fauna is shared with the South-East Asian mainland and the Sundaic islands, but there is relatively little overlap with islands to the north and east. The fauna, especially in the lowland rainforests, is characterised by a remarkable diversity and endemism. Borneo has some of the oldest rainforests in the world with lowland rainforest covering most of the island. These rainforests are amongst the richest in the world with a biodiversity rivalling those of the Amazon and New Guinea.

## Avifauna

The birds of Borneo are essentially Asian in origin, exhibiting many similarities with the Thai-Malay Peninsula and Sumatra, and to a lesser extent Java and Bali. The entire Sundaic region shares the largest proportion of its resident forest birds with Indochina, but Australian, Indian and African influences are also evident. The distribution of birds in Borneo is not uniform and is influenced by habitat, altitude, and geological and climatic history.

# HOW TO USE THIS BOOK

♂

♀

*Jambu*                                    *Bornean Banded Pitta*

In the species accounts section of the book, each bird's scientific name and overall length is given at the beginning of the account, followed by a brief general description that includes the preferred habitat of the species as well as notes on its behaviour as this often provides important clues to identification. Where useful, calls and songs are described. These are notoriously difficult to describe in words and can only be rendered subjectively.

The following terms are used to describe the abundance of species on Borneo: abundant – a conspicuous species that occurs in very large numbers; common – occurs in large numbers; fairly common – occurs in fair to moderate numbers; uncommon – found in small numbers; rare – recorded in very small numbers. Additionally, the terms 'local' or 'locally common/uncommon' are used for species that occur in restricted areas but can be common or uncommon there.

'North Borneo' refers to Sabah, Brunei, Sarawak, northern East Kalimantan, and far-north Central and West Kalimantan, while 'South Borneo' refers to South Kalimantan, and southern East, Central and West Kalimantan. The divide can be thought of as occurring roughly at the equator. Generally, distributions within Borneo are given in detail according to state and province, and divisions within those units (E & W Sabah; Brunei; E, C & W Sarawak; E, C, W & S Kalimantan).

### A note on the photos
All images show the adult form, unless stated otherwise. Please refer to the key below for further information:

| | | | | | | | |
|---|---|---|---|---|---|---|---|
| Male | ♂ | Young | Y | Breeding | Br | Summer | Sum |
| Female | ♀ | Juvenile | Juv | Non-breeding | Nbr | Winter | Win |

# THE HABITATS OF BORNEO

The island of Borneo can be divided into seven ecoregions, that is, units of land that contain distinct assemblages of species, natural communities and environmental conditions. These are: Borneo peat swamp forests; Southern Borneo freshwater swamp forests; Sundaland heath forests; Borneo montane rainforests; Borneo lowland rainforests; Sunda shelf mangroves; and Kinabalu montane alpine meadows.

Three of these ecoregions (Borneo montane rainforests, Borneo lowland rainforests, and Kinabalu montane alpine meadows) are considered globally outstanding in terms of species richness and endemism. The ecoregion with the highest biodiversity is lowland rainforest, and its rate of endemism is correspondingly the highest.

## Peat swamp forests

This forest type is largely absent from Sabah but there are large expanses further south, mostly near the coast, especially in Central Kalimantan. Peat swamp forests form when sediments build up behind mangroves as rivers drain to the coast. They are rain-fed, and the soils are nutrient deficient and acidic with a resultant lower species diversity than that of lowland dipterocarp forest, but a handful of species are commoner in peat swamp forest. Few plant or animal species are endemic to peat swamp forest and there is some overlap with the flora of heath forests. Of the birds, only Javan White-eye and Hook-billed Bulbul are near-endemic to this forest type.

## Southern Borneo freshwater swamp forests

This ecoregion is restricted to Kalimantan, mostly inland of the south-west coast with some smaller areas in the centre. These forests are associated with coastal swamps, inland lakes and rivers, and cover flat and low-lying alluvial plains. The waterlogged forests are characterised by regular flooding and are more diverse than peat swamp forests. Faunal diversity correlates with floristic diversity. Primate densities can be high and up to 360 bird species have been recorded in this forest type including several species of hornbill, but with only one near-endemic, Javan White-eye.

## Sundaland heath forests (kerangas)

Known in Borneo as kerangas, a local Iban name for heath forest (literally 'land which cannot grow rice'), this forest type is characterised by poor soils and low plant diversity. It is found in scattered areas throughout Borneo, but mostly in Kalimantan, and is usually found on plateaux, ridges and raised beaches. It is characterised by well-drained, sandy and acidic soils with a low, uniform and single-layered canopy. The flora is not species-rich, but many species of palm occur, along with many carnivorous plants that grow in response to nitrogen-poor soils. The low plant diversity is reflected by a low diversity of birds, generally less than half that of lowland dipterocarp forests.

## Borneo montane rainforests

The mountains of Borneo are like islands in a sea of lowland dipterocarp forests.

*Proboscis Monkey*

*Burbidge's Picture Plant*

*Plain Pygmy Squirrel*

*Bornean Keeled Pit-viper*

Over the complex geological history of the island, the isolation of these mountains has produced a unique and diverse set of montane species. The Bornean montane rainforests are in the central highlands of the island, above 1,000m elevation. The montane forests on these mountains are cooler than the lowland forests – every 1,000m one travels above sea level sees a mean 5°C drop in temperature. Rainfall and humidity are also higher than in the lowlands. These cooler conditions mean that many plants normally found in temperate regions grow here. In excess of 250 species of resident birds have been recorded and of Borneo's endemic bird species, 36 (69%) are montane. However, species diversity decreases with altitude. Mount (Gunung) Kinabalu is of particular importance, with 64% of Borneo's endemics found here.

*Birding by electric boat on the Kinabatangan River*

## Borneo lowland rainforests

Borneo's stable climatic conditions have given rise to the world's richest assemblage of flowering plants. The lowland rainforests display a marked stratification, and their layers can be roughly categorised as emergent, upper, middle and lower storeys, and forest floor. Commercially valuable dipterocarps dominate the emergent and upper storey. Rainforest birds in particular also show a clear vertical stratification, with several groups specific to the canopy, middle and lower storeys, and the forest floor, respectively. A single family of massive flowering trees, the dipterocarps, dominate this forest. They are so named for their two-winged fruits (actually, many have three to five wings) and they are supported by huge buttresses to stabilise them in the shallow rainforest soils. Forest trees are often adorned with shade-tolerant lianas and epiphytes – ferns and orchids. Cauliflory, meaning stem flower, is a feature of South-East Asian lowland forests – this is where fruits and flowers grow directly on the trunks and larger branches. These types of plant tend to be pollinated by larger animals such as frugivorous bats, arboreal mammals and larger birds.

Lowland rainforests exhibit higher species richness but lower degrees of endemism than montane forests. Avian diversity reaches its peak here. More than 60% of the species on Borneo are confined to this forest type, while almost 80% are dependent on it to some degree. The great structural and floristic diversity combined with the stable conditions exhibited by lowland dipterocarp forest has contributed to this bird diversity.

Borneo's lowland dipterocarp rainforests are the most extensive in the region but they have been much exploited for timber in the last few decades and are rapidly being replaced by plantations, especially oil palm.

## Sunda shelf mangroves

Borneo's mangroves are considered to be amongst the most biologically diverse on Earth. They have a patchy distribution, mostly on coasts. Biodiversity and endemism rates are relatively low. Despite this, they are immensely important as they occur at the interface between terrestrial and marine habitats, stabilising coastlines and providing nurseries for numerous species of fish.

A number of bird species on Borneo are largely dependent on mangroves but few are restricted to this habitat type. This forest type is valuable for passage migrants and as roost sites for a number of species. The remarkable Proboscis Monkey is restricted to mangroves and peat swamp forests, where it feeds on the young leaves of mangrove trees with the aid of its highly specialised digestive system.

## Kinabalu montane alpine meadows

The highest elevations of Mount Kinabalu and the Crocker Range are home to Kinabalu montane alpine meadows, a mix of alpine shrublands and lower-stature forests notable for numerous endemic species, including many orchids. These meadows have been isolated from other mountain chains for many millions of years and are considered to be globally outstanding in terms of species richness and endemism. The slopes of Mount Kinabalu and surrounding mountains support more than 4,500 plant species, including in excess of 750 species of orchids, 600 species of ferns, 13 species of *Nepenthes* pitcher plants (five endemic) and 78 species of figs, 13 of which are endemic to the ecoregion. There is some overlap with this ecoregion and Borneo montane rainforests. Some 326 species of bird have been recorded, 23 of them endemic. One of Borneo's endemics, Friendly Bush Warbler, is confined to Mount Kinabalu and neighbouring Mount Trus Madi.

*A lowland rainforest waterfall*

# WHERE TO WATCH BIRDS ON BORNEO

Due to its stable climate, Borneo can be visited and birded any time outside the rainy season, which is November to January. The rainy season doesn't preclude birding but access to some sites and areas may be limited due to seasonal flooding.

## MOUNT KINABALU, SABAH

Located on the doorstep of Sabah's major city, Kota Kinabalu, Mount Kinabalu is Borneo's highest peak and a major centre of worldwide importance for biodiversity. Mount Kinabalu and its surrounding forests are protected within the World Heritage Kinabalu Park covering a wide range of habitats, from rich tropical lowland and hill rainforest to tropical mountain forest, subalpine forest and scrub on the higher elevations. At 4,095m, Mount Kinabalu is the highest peak between New Guinea and the Himalayas, formed by a granitic intrusion that was thrust upwards from Earth's crust an estimated one million years ago.

Due to the area's great altitudinal and climatic gradient from tropical forest to alpine conditions and its unique geological history, Mount Kinabalu has an exceptionally diverse biota and high endemism. An estimated 6,000 plant species have been recorded within the park boundaries, contributing to its equally impressive faunal diversity. Over 100 species of mammal have been recorded in the park, as well as up to 80 species of amphibians and 112 species of reptiles. The diversity of birds in the park is notable, with at last count 326 species, of which 23 are endemic to the island.

**Areas to explore:** Power Station Road, Silau Silau Trail, Liwagu River Trail and Mempening Trail.

**Birding highlights:** Crimson-headed Partridge, Red-breasted Partridge, Golden-naped Barbet, Mountain Barbet, Whitehead's Trogon, Whitehead's Broadbill, Bornean Green Magpie, Bornean Treepie, Bornean Stubtail, Mountain Wren-Babbler,

*Mount Kinabalu*

*Mount Kinabalu at sunrise*

Chestnut-hooded Laughingthrush, Sunda Laughingthrush, Bare-headed Laughingthrush, Eyebrowed Jungle-Flycatcher, Mountain Black-eye, Whitehead's Spiderhunter, Temminck's Sunbird and Black-sided Flowerpecker.

**Access:** Kinabalu Park can be reached by car or public bus, a drive of roughly one and a half hours from Kota Kinabalu. Accommodation is available within the park and outside the park entrance, at the main headquarters and at Poring, as well as in the nearby towns of Kundasang and Ranau.

### PORING HOT SPRINGS, SABAH

Poring is located within the eastern boundary of Kinabalu Park and is situated 40km from the park headquarters. During World War II the Japanese developed the area for its hot sulphur spring baths. The name 'Poring' comes from the indigenous Kadazan-Dusun word for a bamboo species found in the area. Poring is about 400m above sea level with excellent stands of hill forest; as a result a different range of bird species can be found here. A 43m-high walkway allows canopy-dwelling birds to be observed at eye level. Another main attraction at Poring is the Rafflesia, the world's largest flower.

**Areas to explore:** Kipungit Waterfall (30-minute walk), Laganan Waterfall (90-minute walk) and Poring Canopy Walkway.

**Birding highlights:** Rufous-collared Kingfisher, Red-throated Barbet, Gold-whiskered Barbet, Black-and-yellow Broadbill, Blue-banded Pitta, Crested Shrikejay, Scaly-breasted Bulbul, Yellow-bellied Warbler, Fulvous-chested Jungle-Flycatcher, Orange-bellied Flowerpecker and Bornean Spiderhunter.

**Access:** Private car or public bus from Kota Kinabalu or Kinabalu Park headquarters.

*Kota Kinabalu sunset*

*Crocker Range at sunset*

## TAMBUNAN, CROCKER RANGE NATIONAL PARK, SABAH

The journey by road from Kota Kinabalu to the Rafflesia Information Centre takes nearly two hours, with roads winding up and down the mountainous Crocker Range. The centre protects an area of forest where the Rafflesia plants grow but birders will want to concentrate their efforts along the road that bisects excellent hill forest at an average elevation of 1,200m. Fruiting trees can be particularly productive for some of the special birds in this area. With less passing traffic, early mornings will be the most productive for wildlife.

**Birding highlights:** Mountain Serpent-Eagle, Bornean Barbet, Mountain Barbet, Long-tailed Broadbill, Pygmy White-eye, Fruit-hunter, Bornean Bulbul, Bornean Leafbird and Whitehead's Spiderhunter.

**Access:** Private car from Kota Kinabalu. Accommodation is not readily available in this area so a day trip or birding en route to Mount Kinabalu are recommended.

## SEPILOK, SABAH

The Sepilok Orang Utan Rehabilitation Centre is a major tourist attraction, but the centre also protects a large expanse of tropical lowland rainforest. Exploration of the many trails and boardwalks can be very productive for birding. The nearby Rainforest Discovery Centre features an extensive network of trails as well as long canopy walkways connected by four towers that allow some of the best canopy birding on Borneo.

**Areas to explore:** Orang Utan Rehabilitation Centre and Rainforest Discovery Centre.

*Crocker Range*

**Birding highlights:** Jambu Fruit-Dove, Raffles's Malkoha, Banded Bay Cuckoo, Square-tailed Drongo-Cuckoo, Grey-rumped Treeswift, Wallace's Hawk-Eagle, Rufous-bellied Eagle, Oriental Bay Owl, Barred Eagle-Owl, Red-naped Trogon, Bushy-crested Hornbill, Black Hornbill, Oriental Dwarf-Kingfisher, Banded Kingfisher, Rufous-collared Kingfisher, Brown Barbet, Blue-eared Barbet, Malaysian Honeyguide, Banded Woodpecker, White-fronted Falconet, Blue-rumped Parrot, Blue-naped Parrot, Black-and-yellow Broadbill, Black-crowned Pitta, Bornean Bristlehead, Green Iora, Bronzed Drongo, Black-naped Monarch, Slender-billed Crow, Bold-striped Tit-Babbler, Common Hill Myna, Orange-bellied Flowerpecker, Ruby-cheeked Sunbird, and Asian Fairy-bluebird.

**Access:** Sepilok can be reached from the east coast town of Sandakan by private car, taxi or public bus. The drive takes about an hour. There is a range of accommodation options in the vicinity of both centres.

## KINABATANGAN RIVER & GOMANTONG CAVES, SABAH

The Kinabatangan River is the second longest in Malaysia and the eighth longest on the island of Borneo. Since 1997, 270km$^2$ of the lower Kinabatangan floodplain is protected in the Kinabatangan Wildlife Sanctuary. The sanctuary protects a mosaic of riparian swamp forest, oxbow lakes, nipah palm swamp and mangroves, which support a wide variety of mammal and birdlife, one of the richest ecosystems on Earth. Large areas are periodically flooded during the rainy season.

This centre of biodiversity is home to over 290 bird and 60 mammal species, including threatened Bornean Orang Utan, Bornean Pygmy Elephant, Estuarine Crocodile and Irrawaddy Dolphin. It is also one of only two areas in the world inhabited by ten species of primate, four of which are endemic to Borneo. All eight species of hornbill occurring on Borneo can be seen here, too.

The Gomantong Caves are a major attraction for wildlife watchers. This huge network of limestone caves is the best place

*Kinabatangan River*

to see nesting Edible-nest, Black-nest and Mossy-nest Swiftlets. At dusk thousands of insectivorous bats can be seen streaming out of the caves, with Bat Hawks and Peregrine Falcons in rapid pursuit.

**Areas to explore:** Lodges catering for everyone from the budget traveller to the luxury tourist can be found in the Sukau, Bilit and Abai areas. All these lodges offer packages that include transport to the sanctuary and surrounds, guided river cruises, often in boats with electric motors, and meals. Additional tours such as night walks or night cruises to spot crocodiles, birds and nocturnal animals are usually available as well.

**Birding highlights:** Little Green Pigeon, Red-billed Malkoha, Chestnut-breasted Malkoha, Violet Cuckoo, Moustached Hawk-Cuckoo, Brown-backed Needletail, Plume-toed Swiftlet, Mossy-nest Swiftlet, Black-nest Swiftlet, White-nest Swiftlet, Storm's Stork, Lesser Adjutant, Oriental Darter, Jerdon's Baza, Bat Hawk, Wallace's Hawk-Eagle, Buffy Fish-Owl, White-crowned Hornbill, Rhinoceros Hornbill, Bushy-crested Hornbill, Black Hornbill, Wrinkled Hornbill, Blue-eared Kingfisher, Rufous-backed Dwarf-Kingfisher, Stork-billed Kingfisher, Grey-and-buff Woodpecker, Great Slaty Woodpecker, White-fronted Falconet, Long-tailed Parakeet, Black-and-red Broadbill, Hooded Pitta, Fiery Minivet, Bornean Bristlehead and Black-throated Babbler.

**Access:** The Kinabatangan Wildlife Sanctuary is usually reached from Sandakan to the north, either overland in private vehicles or lodge shuttle buses, or by boat via the Kinabatangan River mouth.

## DANUM VALLEY CONSERVATION AREA, SABAH

Danum Valley is 438km$^2$ area of relatively undisturbed lowland dipterocarp forest and is arguably Borneo's – if not South-East Asia's – premier wildlife location. The valley is bowl-shaped, with a maximum elevation of 1,093m. Before it became a conservation area there were no human settlements, so hunting, logging and other human interference was virtually non-existent. Ten species of primate occur, including the largest, Bornean Orang Utan, and the smallest, Western Tarsier. Over 320 species of bird have been recorded in this remarkable expanse of rainforest. The diversity of flora and fauna in Danum Valley has led it to be considered one of the world's most complex ecosystems.

Early morning is the best time to enjoy the wildlife, when the birds are active

*Birdwatching in the Danum Valley, Sabah*

and the temperatures are relatively cool. Canopy walkways at various locations allow excellent wildlife-viewing opportunities.

**Areas to explore:** There are few lodges located within the conservation area, and stays at all of them must be prearranged. The Danum Valley Field Centre is a research establishment for scientific and educational purposes. Previously reserved for researchers, the facilities are now available for visitors looking for a basic stay while exploring Danum Valley. The Borneo Rainforest Lodge, located on the banks of the Danum River, is an award-winning eco-lodge with very comfortable chalets connected by boardwalks to a central dining and reception area. Visitors can enjoy guided walks on rainforest trails and the canopy walkway in the early morning and at dusk, as well as night safari walks or drives. The Kawag Nature Lodge is located in buffer zone II of the Danum Valley Conservation Area and offers a slightly less luxurious option with similar opportunities to explore the area.

**Birding highlights:** Crested Partridge, Scaly-breasted Partridge, Great Argus, Crested Fireback, Thick-billed Green Pigeon, Green Imperial-Pigeon, Jambu Fruit-Dove, Large Frogmouth, Gould's Frogmouth, Whiskered Treeswift, Silver-rumped Spinetail, Bornean Ground-Cuckoo, Chestnut-bellied Malkoha, Violet Cuckoo, Great-billed Heron, Crested Serpent-Eagle, Black Eagle, Crested Goshawk, Brown Wood-Owl, Scarlet-rumped Trogon, Diard's Trogon, Red-naped Trogon, Helmeted Hornbill, Wreathed Hornbill, Rufous Piculet, Rufous Woodpecker, Buff-rumped Woodpecker, Buff-necked Woodpecker,

*Lowland rainforest in the Danum Valley, Sabah*

Crimson-winged Woodpecker, White-bellied Woodpecker, Orange-backed Woodpecker, Gold-whiskered Barbet, Red-crowned Barbet, Red-throated Barbet, Yellow-crowned Barbet, Blue-eared Barbet, Brown Barbet, Red-bearded Bee-eater, Blue-banded Kingfisher, Rufous-collared Kingfisher, White-fronted Falconet, Blue-crowned Hanging Parrot, Giant Pitta, Bornean Banded Pitta, Blue-headed Pitta, Banded Broadbill, Green Broadbill, Scarlet Minivet, Sunda Cuckooshrike, Dark-throated Oriole, Black-winged Flycatcher-shrike, Large Woodshrike, Rufous-winged Philentoma, Maroon-breasted Philentoma, Bornean Bristlehead, Greater Racket-tailed Drongo, Spotted Fantail, Crested Shrikejay, Blyth's Paradise-flycatcher, Yellow-rumped Flowerpecker, Thick-billed Spiderhunter, Long-billed Spiderhunter, Little Spiderhunter, Purple-naped Spiderhunter, Crimson Sunbird, Asian Fairy-bluebird, Greater Green Leafbird, Dusky Munia, Grey-headed Canary-Flycatcher, Yellow-bellied Prinia, Rufous-tailed Tailorbird, Hairy-backed Bulbul, Straw-headed Bulbul,

Chestnut-backed Scimitar-Babbler, Chestnut-rumped Babbler, Fluffy-backed Tit-Babbler, Chestnut-winged Babbler, Black-capped Babbler, Short-tailed Babbler, Striped Wren-Babbler, Black-throated Wren-Babbler, Bornean Wren-Babbler, White-crowned Shama, Bornean Blue Flycatcher, White-crowned Forktail and Rufous-chested Flycatcher.

**Access:** Located two hours' drive, around 80km, from the town of Lahad Datu. The lodges can only be accessed by shuttle buses operated by the lodges and the research centre.

## TAWAU HILLS PARK, SABAH

Tawau Hills Park is an area of 280km² of lowland dipterocarp rainforest surrounded by oil palm and cacao plantations. It was established in 1979, primarily to protect the water catchment area of Tawau town, Sabah. Numerous trails allow access to wildlife watching away from the often-busy picnic area adjacent to the headquarters.

**Birding highlights:** Great Argus, Barred Eagle-Owl, Scarlet-rumped Trogon, Diard's Trogon, White-crowned Hornbill, Helmeted Hornbill, Rhinoceros Hornbill, Yellow-crowned Barbet, Gold-whiskered Barbet, Crimson-winged Woodpecker, Blue-headed Pitta, Black-and-yellow Broadbill, Grey-cheeked Bulbul, Finsch's Bulbul, Chestnut-backed Scimitar-babbler, Grey-headed Babbler, White-necked Babbler, Sooty-capped Babbler, Brown Fulvetta, Rufous-tailed Shama, Greater Green Leafbird, Yellow-rumped Flowerpecker and Streaky-breasted Spiderhunter.

**Access:** The park offers picnic areas, camping sites, and chalets. It can be reached by private vehicle or taxi from Tawau, 28km to the north-west of the park.

## TABIN WILDLIFE RESERVE, SABAH

Tabin Wildlife Reserve is located in eastern Sabah. It is one of the largest wildlife reserves in Malaysia, comprising an area of approximately 1,200km² in the centre of the Dent Peninsula, north-east of Lahad Datu town, south of the lower reaches of the Segama River and north of the Silabukan Forest Reserve. The three largest mammals of Sabah, namely Borneo Pygmy Elephant, Banteng and Sumatran Rhinoceros (now extinct) have been recorded within the reserve, as well as nine species of primate and three species of cat. More than 300 species of bird can be found here. One of the

*Sungai Menanggol, a tributary of the Kinabatangan*

highlights of Tabin are the active and mineral-rich mud volcanoes, which attract wildlife for their mineral content, allowing excellent wildlife and birding opportunities.

**Birding highlights:** Chestnut-necklaced Partridge, White-fronted Falconet, Jerdon's Baza, Rufous-bellied Eagle, Large Green Pigeon, Green Imperial Pigeon, Long-tailed Parakeet, Moustached Hawk-Cuckoo, Violet Cuckoo, Little Bronze Cuckoo, Brown Wood Owl, Red-bearded Bee-eater, Wreathed Hornbill, White-bellied Woodpecker, Green Broadbill, Dusky Broadbill, Fiery Minivet, Dark-throated Oriole, Maroon-breasted Philentoma, White-breasted Woodswallow, Bornean Black Magpie, Crimson Sunbird, Asian Fairy-bluebird, Common Hill Myna, Malaysian Blue Flycatcher, Verditer Flycatcher, Fluffy-backed Tit-Babbler, Black-capped Babbler and Striped Wren-Babbler.

**Access:** Tabin is four hours' drive from Lahad Datu by shuttle bus when accommodation is prearranged in the Tabin Wildlife Reserve.

## BAKO NATIONAL PARK, SARAWAK
This small park of only 27km² is located on the northern tip of the Muara Tebas peninsula, about 40km from Sarawak's capital, Kuching. Despite its small size, Bako protects a diverse range of habitats including swamp forest, scrub-like padang vegetation, mangrove forest, delicate cliff vegetation and lowland rainforest. This beautiful area features vistas of steep sea cliffs, rocky headlands and stretches of white, sandy bays. It is one of the best places to see the unique Proboscis Monkey, which is

*Bako National Park*

endemic to Borneo. Over 250 species of bird have been recorded here.

**Areas to explore:** There is an excellent network of marked walking trails of different lengths, allowing visitors access. In addition, various beaches are accessible by boat.

**Birding highlights:** Pink-necked Green Pigeon, Eurasian Whimbrel, White-bellied Sea-Eagle, Red-crowned Barbet, Pied Triller, Black-winged Flycatcher-shrike, Greater Racket-tailed Drongo, Rufous-tailed Tailorbird, Bold-striped Tit-Babbler, White-chested Babbler, Velvet-fronted Nuthatch, Mangrove Blue Flycatcher, Ruby-cheeked Sunbird and Asian Fairy-bluebird.

**Access:** The park can only be reached by a 20-minute boat ride from the village of Kampung Bako. It is often visited as a day trip from Kuching, though accommodation (campground and forestry service bungalows) is available.

## PAYA MAGA, SARAWAK
In the far north of Sarawak in the Ulu Trusan region, the mountainous Paya Maga IBA is one of the state's most promising new birding destinations. It can be reached by 4WD vehicle from the town of Lawas,

*En route to Paya Maga*

and a campsite in the forest at the higher elevations allows access to some very special attractions, including Bornean Frogmouth, Bornean Leafbird, Bare-headed Laughingthrush and the recently rediscovered Black Oriole.

**Birding highlights:** Blue-banded Pitta, Bornean Banded Pitta, Black Oriole, Crested Shrikejay, Malaysian Rail-Babbler, Eyebrowed Wren-Babbler, Scaly-breasted Bulbul, Bornean Bulbul and Pygmy White-eye.

**Access:** Paya Maga can visited by prearranged tours with speciality birding outfits or by hiring a 4WD vehicle in Lawas. Accommodation is a basic forest camp with simple cabins. Guides and porters can be hired in the nearby village of Long Tuyo. All bedding and food must be carried in.

### BAKELALAN, SARAWAK
The village of Bakelalan allows access to Pulong Tau National Park, Sarawak's largest reserve. The reserve covers

nearly 600km² of pristine montane rainforest, occupying the western flank of the Kelabit highlands of north-east Sarawak. Its spectacular mountain landscapes include Mount Murud, Sarawak's highest peak (2,424m) and the Tama Abu range. The whole area is home to the ethnic Lun Bawang people, meaning 'people of the land'.

More than 300 species of birds are listed for this park. With a broad altitudinal range and habitat diversity, Pulong Tau National Park is home to numerous endemic birds, including Bornean Whistler, Mountain Barbet, Whitehead's Broadbill and the enigmatic Dulit Frogmouth.

**Birding highlights:** Bornean Frogmouth, Blyth's Hawk-Eagle, Mountain Serpent-Eagle, Black Eagle, Black-thighed Falconet, Bornean Barbet, Mountain Barbet, Hose's Broadbill, Long-tailed Broadbill, Grey-headed Babbler, Rufous-chested Flycatcher, Yellow-rumped Flowerpecker, Bornean Leafbird, Bornean Spiderhunter and Whitehead's Spiderhunter.

**Access:** Bakelalan is serviced by MASWings with regular flights from Lawas; alternatively it can be reached by road, requiring a 4WD vehicle. This drive can take anywhere from five to nine

*Bakelalan*

hours, depending on the road conditions. It is a three- to four-hour drive from Paya Maga. Homestay accommodation is available but should be prearranged.

## BORNEO HIGHLANDS RESORT, SARAWAK

Borneo Highlands Resort is located in the lush green rainforest of the Penrissen Range at a cool 1,000m above sea level. The mountain range straddles the rugged border of Sarawak (Malaysian Borneo) and Kalimantan (Indonesian Borneo) and is home to a remarkably diverse avifauna.

**Areas to explore:** The main access road, trails and the road to the Kalimantan Viewpoint on the border with Malaysia and Indonesia.

**Birding highlights:** Red-billed Malkoha, Mountain Serpent-Eagle, Blyth's Hawk-Eagle, Barred Eagle-Owl, Red-bearded Bee-eater, Bornean Barbet, Olive-backed Woodpecker, Crested Shrikejay, Temminck's Babbler and Grey-breasted Spiderhunter.

**Access:** Located only 60km south of Kuching, the Borneo Highlands can be reached by taxi or private vehicle in roughly an hour and a half. Accommodation is available at the Borneo Highlands Resort.

## KUBAH, SARAWAK

This national park, located only half an hour from Kuching, protects a small area of only 22.3km² of mixed dipterocarp forest. It sits on a sandstone plateau, part of the Matang Range comprising three peaks, the highest being Mount Serapi at 900m. Over 250 species of bird, as well as mammals such as Bearded Pig and Mouse Deer, and many species of amphibian and reptile occur.

**Areas to explore:** The road to Serapi Peak, Waterfall Trail, Matang Wildlife Centre and Rayu Trail.

**Birding highlights:** Raffles's Malkoha, Indian Cuckoo, Malaysian Eared-nightjar, Silver-rumped Needletail, Crested Serpent-Eagle, Red-naped Trogon, Scarlet-rumped Trogon, Banded Kingfisher, Red-crowned Barbet, Rufous Piculet, Crimson-winged Woodpecker, Blue-crowned Hanging Parrot, Green Broadbill, Black-and-yellow Broadbill, Blue-banded Pitta, Scarlet Minivet, Lesser Cuckooshrike, Dark-throated Oriole, Black-winged Flycatcher-shrike, Rufous-winged Philentoma, Spotted Fantail, Crested Shrikejay, Black Magpie, Grey-bellied Bulbul, Scaly-breasted Bulbul, Rufous-fronted Babbler, Bornean Wren-Babbler, Chestnut-backed Scimitar-Babbler, White-necked Babbler, Moustached Babbler, Bornean Blue Flycatcher, Grey-chested Jungle-Flycatcher, Verditer Flycatcher, Chestnut-naped Forktail, Yellow-breasted Flowerpecker, Scarlet-breasted Flowerpecker, Ruby-cheeked Sunbird, Plain Sunbird, Brown-throated Sunbird, Purple-naped Spiderhunter, Little Spiderhunter, Asian Fairy-bluebird and Lesser Green Leafbird.

**Access:** Less than an hour's drive from Kuching by taxi or private vehicle. A range of accommodation is available within the park. Cabins or hostels can be booked via the Sarawak National Park website.

## ULU TEMBURONG NATIONAL PARK, BRUNEI

This park is located in eastern Brunei and encompasses an area of 550km² of lowland rainforest. This remote area can only be reached by river from

Brunei's capital, Bandar Seri Begawan. The journey begins with a water taxi trip through the mangrove forests of the Limbang River delta and continues with a longboat trip up the Temburong River. Ulu Temburong's forests are exceptionally rich in flora and fauna, with over 200 species of bird and mammals such as Bornean Orang Utan, Bornean Gibbon and the world's smallest squirrel, Plain Pygmy Squirrel.

**Areas to explore:** The park boasts an extensive network of boardwalks, bridges and stairways, as well as a 50m-high canopy walkway, which provides a wonderful view over the forest canopy.

**Access:** The Ulu Ulu Resort is just north of the park on the Temburong River and can be booked through travel agents in Bandar Seri Begawan.

### TANJUNG PUTING NATIONAL PARK, KALIMANTAN

This Indonesian national park is located in the south-east part of the Indonesian province of Central Kalimantan, west of the town of Kumai. It protects over 4,160km² of dry dipterocarp forest, peat swamp forest, heath forest, mangrove and coastal beach forest, as well as secondary forest. Bornean Orang Utan and Proboscis Monkey are two of the more spectacular mammals found here; other species include Bornean Gibbon, Long-tailed Macaque, Horsfield's Tarsier and Clouded Leopard. Almost 200 species of bird have been recorded but many new birds await to be discovered in this under-appreciated area.

**Birding highlights:** Black Partridge, Crestless Fireback, Storm's Stork, Great-billed Heron, Large Green Pigeon, Stork-billed Kingfisher, Red-crowned Barbet, Long-tailed Parakeet, Black-and-red Broadbill, Mangrove Whistler, Bornean Bristlehead and Javan White-eye.

**Access:** Visits to Tanjung Puting start at Pangkalan Bun village, boarding a wooden boat called a *klotok*, typically for a three-night live-aboard trip. The boats are simple but comfortable with a top-deck dining area, small bedrooms and shared bathroom facilities. There are also numerous lodges located outside the park.

### SUNGAI WAIN, KALIMANTAN

The Sungai Wain Protected Forest is the last intact area of primary old-growth lowland rainforest in the region. The relatively small area of only 100km² belies the importance of the reserve, as it protects significant populations of three Endangered endemic species: Bornean Peacock-Pheasant, Bornean Bay Cat and Bornean Gibbon. It also supplies a large portion of the freshwater needs of Borneo's fourth-largest city, Balikpapan.

**Areas to explore:** A single main trail runs through the centre of the reserve, with numerous smaller side trails.

**Birding highlights:** Bornean Peacock-Pheasant, Large Frogmouth, Lesser Adjutant, Reddish Scops-Owl, Diard's Trogon, Orange-backed Woodpecker, Garnet Pitta, Black-throated Babbler, Grey-breasted Babbler and Rufous-tailed Shama.

**Access:** About one hour by vehicle from Balikpapan. Accommodation at homestays is available outside the park entrance. Permits must be obtained to enter the park.

# SPECIES ACCOUNTS

## Wandering Whistling-Duck *Dendrocygna arcuata* 40–45cm

A medium-sized, mostly brown duck. The paler head and neck contrast with the dark brown back and wings and chestnut underparts. The distinctive, fluffy white plumes on the flanks are diagnostic. In flight the long neck is carried lower than the body.

**Where to see** A common resident in paddyfields, rivers, lakes and wetlands in Sabah but scarcer in Kalimantan and Sarawak.

## Garganey *Anas querquedula* 37–41cm

The male is very distinctive, streaky brown overall with a broad white eye-stripe and elongated feathers on the wings. The females and non-breeding males are less distinctive, but also show a buffish eye-stripe as well as a cheek-stripe.

**Where to see** The breeding male plumage is less likely to be seen on Borneo as this is an uncommon non-breeding visitor that may be found in lowland wetlands throughout the island.

♀ Nbr

# Philippine Scrubfowl *Megapodius cumingii* 32–38cm

A chicken-like bird with a dark grey body, brown wings and back and a small, brown crest. Bare reddish-pink skin around the sides of the head and very large, strong feet. These birds favour beach forest and coastal scrub where they feed mostly on invertebrates. The females lay their eggs in mounds of decaying vegetation that provide warmth to the developing embryos. Apart from tending the mounds to regulate temperature, the adults offer no parental care, and the chicks are entirely independent when they hatch.

**Where to see** Offshore islands in the north.

# Red-breasted Partridge   *Arborophila hyperythra*   25cm

A small, dumpy bird that frequents the forest floor where it feeds on insects, fruits and seeds. Secretive but vocal – more often heard than seen, the call is a strident, repetitive duetting *pyu-pyu-pyu-pyu...* lasting about 20 seconds and becoming steadily faster and louder. Upperparts are olive-brown with bold black spots and large white spots on the flanks. The head and underparts are rusty-red, with a prominent dark eye-line.

**Where to see** Endemic, in the north-central mountains.

Juv

Ad

**Pheasants, Grouse and Allies**

# Crested Partridge  *Rollulous rouloul*  26cm

A remarkable looking ground-dwelling partridge. Males and females look very different. The male is dark glossy blue with a dark green rump and a spectacular reddish upright crest, while the female is grass-green with a grey head. Both have red legs and a red patch of skin around the eye. They are secretive birds of lowland and hill forests, where they forage for fruits, seeds and invertebrates in groups of up to a dozen.

**Where to see** Throughout the island.

# Bornean Crested Fireback  *Lophura nobilis*  66cm (male); 56cm (female)

This spectacular pheasant is found in groups, usually consisting of a male with 4–5 females and immatures. They forage for fruits, seeds and invertebrates on the forest floor. The male has dark glossy blue and chestnut-brown underparts; the plumed crest, bright blue facial skin, bright buffy tail feathers and bright coppery-red rump are distinctive. The female is chestnut with broad white fringes on the underparts.

**Where to see** Lowland forests throughout the island.

# Crimson-headed Partridge   *Haematortyx sanguiniceps*  25cm

A very unusual, small, endemic partridge of the hills and mountains. Usually found in small coveys of two to five birds. The dark brown body is topped with a blood-red head. The male has a yellow bill, while the similar-looking female has a dark bill. The song is a loud, ringing, repeated two-note *kong-krrang kong-krrang...* often performed as a duet.

**Where to see** Endemic, Mount Kinabalu and the Crocker Range.

# Great Argus   *Argusianus argus*   165cm (male); 60cm (female)

This large and long-bodied pheasant is unmistakable. The male is dark brown with black and white vermiculations, bare blue facial skin and a very long tail – the longest tail feathers of any bird. The female is smaller and darker with a shorter tail. Usually solitary but the male is sometimes seen in company with two or three females. The male is very vocal and can be heard giving a loud, explosive *ka-wow* call at all times of day. Found in hilly areas of primary evergreen forests but populations are declining.

**Where to see** Throughout Borneo but most often seen in protected areas of Sabah.

# Spotted Dove *Streptopelia chinensis* 32cm

This common resident is a medium-sized, pale grey, slender dove with a conspicuous, white-spotted black collar, pinkish-grey breast and broad white outer tail tips, noticeable in flight. It is usually found in secondary forest, open woodlands, cultivated areas, parks and gardens. Mostly feeds on seeds, foraging on ground. The call is a mellow, repeated *ku koo-koor*. The flight is often noisy with loud wing-clapping, and the tail is spread to display the white tail tips.

**Where to see** Throughout Borneo in all modified habitats.

## Asian Emerald Dove
*Chalcophaps indica* 25cm

A widespread ground-dwelling dove characterised by an iridescent bronze-green back and wings, with purplish head and underparts and a pale cap. They favour lowland forests and are tolerant of secondary growth. They forage unobtrusively on the ground, often in pairs, for seeds, fallen fruits and invertebrates. When disturbed they will flush noisily in a low, direct and very fast flight. More commonly heard than seen, the song is a low-pitched, mournful *wup woooo*.

**Where to see** Locally common throughout Borneo.

Pigeons and Doves

# Little Cuckoo-Dove *Macropygia ruficeps* 28cm

A locally common, all-rufous dove, with a long tail and a small patch of pale blue skin around the eye. More often heard than seen, the voice is a characteristic, buoyant *wook wook wook,* which is rapidly repeated. It favours lower mountain forests where it feeds in the canopy on fruits and seeds. In flight it resembles a cuckoo or a small hawk with powerful, direct wingbeats.

**Where to see** Hill forests throughout Borneo.

# Zebra Dove *Geopelia striata* 21cm

This small dove is introduced into Borneo, where populations have become established in various cities, probably from cage bird escapees. The plumage lives up to the bird's name, with fine grey and black bars on the back and sides and a pale bluish-grey face and throat. The iris is white with a blue eye-ring.

**Where to see** Locally common in populated areas throughout. Numbers are probably increasing.

## Little Green Pigeon
*Treron olax* 20cm

The smallest of the *Treron* pigeons. The males have a maroon back and wings with a burnt-orange breast, while the females are all green with paler underparts. These pigeons are usually seen in small flocks flying high over the forest, where they feed in fruiting trees, particularly favouring figs. The distinctive but subtle calls are a peculiar, undulating and strangulated *wii...oo...wii...*

**Where to see** Lowland to hill forests, as well as plantations, parks and gardens throughout Borneo.

**Pigeons and Doves**

# Pink-necked Green Pigeon *Treron vernans* 25cm

The male has a burnt-orange breast, green back and wings with yellow edging, and a distinctive pink collar. The female is much plainer, mostly olive-green, with yellowish-green underparts. They tend to be found in more disturbed or open areas than the other *Treron* pigeons, favouring the edges of lowland forests, mangroves, plantations and gardens. They are usually seen in small flocks, feeding in fruiting trees.

**Where to see** Common resident throughout.

# Green Imperial-Pigeon
*Ducula aenea* 48cm

One of the largest pigeons on Borneo. The base colour is grey with a metallic bronze-green back and wings. They are seen singly or in small groups in lowland forests and plantations but may congregate in large numbers in the canopies of fruiting trees. They can sometimes be seen performing a 'roller coaster' display flight in which the bird flies up steeply before stalling, then suddenly entering into a steep dive.

**Where to see** Commonly seen in lowland areas throughout the island.

# Thick-billed Green Pigeon  *Treron curvirostra*  25–31cm

The thick bill is difficult to discern in the field but the distinguishing red cere (skin at the base of the bill) and the broad greenish-blue eye-ring are conspicuous. The male has a maroon back and wings, yellow fringing on the tertials and white scalloping on the undertail-coverts. The female is similar, but the back and wings are green.

**Where to see** Common throughout, in all types of lowland forests.

# Jambu Fruit-Dove
*Ramphiculus jambu*  27cm

This small, colourful pigeon is one of the most beautiful in the region. The base colour is emerald-green and the male has a plum-pink face and throat with white underparts and pink wash on the breast. The female is all green with a dark purple face. This bird is surprisingly hard to spot in the lowland forests that it favours, where it is nomadic in response to fruiting events. Shy and usually solitary or in pairs.

**Where to see** A locally uncommon resident throughout.

# Mountain Imperial-Pigeon
*Ducula badia* 51cm

The largest pigeon on Borneo. This robust bird is grey with a deep purple back and wings, and black tail with a grey terminal band. The bill is dark red with a grey tip and the thick, dark red eye-ring is conspicuous. Essentially a montane dweller, it does disperse widely, including to the lowlands, in search of food. The deep, three-note *ga-woo-woo* calls are diagnostic.

**Where to see** Locally common resident throughout.

# Dark Hawk-Cuckoo
*Hierococcyx bocki* 32cm

This large, parasitic cuckoo resembles a hawk in flight. The head is grey with a yellow eye-ring, the back and wings are brown, the breast is rufous with brown streaks, and the underparts are barred black and white. The sexes are alike. This solitary species utters a mellow, two-note *pui pee-ha*, often repeated incessantly, and is often referred to as a 'brain fever bird'.

**Where to see** An uncommon resident restricted to the north-central mountain ranges and the Meratus Mountains of S Kalimantan.

# Moustached Hawk-Cuckoo
*Hierococcyx vagans*   26cm

A broad black moustachial stripe gives this bird its common name. The cheek and throat are whitish and the upperparts dark brown with faint rufous barring. The underparts are white with clean dark streaks. This shy, solitary cuckoo is much more often heard than seen. The song is a monotonous, melancholy two-note *kang-koh*.

**Where to see** A lowland specialist found throughout the island.

# Indian Cuckoo
*Cuculus micropterus*   32cm

This cuckoo has a grey head with a yellow eye-ring, brownish-grey upperparts and black-and-white barred underparts. The female is similar but has narrower barring on the underparts and a brownish breast. They are generally solitary and feed on insects and small fruits in the canopy. The loud, fluty *po po pa pyo* call is said to fit the mnemonic 'one more bottle'.

**Where to see** Locally common throughout lowland and hill forests up to 1,200m.

# Banded Bay Cuckoo
*Cacomantis sonneratii* 22cm

This heavily banded cuckoo can easily be mistaken for the hepatic (rufous) females or juveniles of a number of other species. The upperparts are finely barred rufous and brown, while the barred underparts are more whitish. It is best distinguished from similarly patterned forms of other species by the better defined facial markings. The distinctive 'smoke yer pepper' call serves as a useful identification feature.

**Where to see** A lowland species throughout the island.

# Plaintive Cuckoo
*Cacomantis merulinus* 23cm

This very widespread Asian species of parasitic cuckoo has a grey head, breast and back, and pinkish-rufous underparts. The adult female is usually finely barred on the underparts, while the hepatic form, along with the juvenile, may easily be mistaken for Banded Bay Cuckoo (see above). The name comes from the bird's call, a plaintive series of descending notes, accelerating and tapering off towards the end.

**Where to see** Common resident throughout.

# Square-tailed Drongo-Cuckoo *Surniculus lugubris* 24cm

This remarkable cuckoo mimics small drongo species, presumably in order to gain protection from mobbing by other bird species, as drongos are extremely aggressive to larger birds. The slender, decurved bill distinguishes it from the drongos, which have heavier and thicker bills. It has a characteristic call consisting of a series of six or seven loud, descending notes.

**Where to see** A fairly common but secretive resident.

# Violet Cuckoo *Chrysococcyx xanthorhynchus*  16cm

In poor light it is difficult to discern the beauty of this species – the glossy violet-purple breast and upperparts are best appreciated in full sunlight. The belly is barred whitish and blackish-purple. The female is less fancy, with a bronze-green crown and back, and barring on the face and underparts. They often fly high over the canopy, uttering their surprisingly far-carrying and distinctive *ki-wiT* calls.

**Where to see** An uncommon resident found throughout the island.

♂

♀

# Red-billed Malkoha
*Zanclostomus javanicus* 42cm

A striking looking malkoha with a grey head, glossy blue back and wings, and rufous underparts. The long, glossy dark blue tail is tipped with white. There is a broad area of bare blue skin around the brown iris and the bill is red. They behave in much the same way as all malkohas – creeping through dense vegetation in the midstorey in search of invertebrate prey.

**Where to see** Found throughout the island in all types of lowland forest.

# Chestnut-breasted Malkoha
*Phaenicophaeus curvirostris* 49cm

The largest malkoha is unmistakable, with a dark grey head, and a glossy dark green back, wings and tail. The tail has a broad chestnut terminal band and unlike all the other malkohas on the island, lacks white tips. The underparts are a rich rufous colour. The pale blue iris is surrounded by a broad patch of crimson bare skin, while the heavy, bicoloured bill is green with a red lower mandible.

**Where to see** Common resident throughout, in all types of forest.

# Raffles's Malkoha   *Rhinortha chlorophaea*   32cm

This is the smallest of the malkohas, a group of large, non-parasitic cuckoos found throughout tropical Asia. This species is unusual in that the males and females differ markedly – the male is all rufous apart from the tail, which is black with fine grey barring and broad white tips, while the female is mostly grey with a rufous back and tail. Its vocalisation is unusual, consisting of a cat-like, mournful *kyar kyar-kyar-kyar-kyar.*

**Where to see** Usually found in pairs in all types of forest throughout Borneo.

# Black-bellied Malkoha   *Phaenicophaeus diardi*   38cm

A mostly green malkoha with a greyish head, an extensive, velvety red bare facial skin patch around the eye, pale green bill, and white tail tips. As the name suggests, the belly and undertail coverts are black. As they secretively clamber around dense vegetation in the forest midstorey, they feed on insects and other invertebrates. The call is a soft, frog-like *kwauk,* which can easily be overlooked.

**Where to see** Can be seen in lowland forests throughout the island.

# Bornean Ground-Cuckoo *Carpococcyx radiatus* 60cm

One of Borneo's most exciting and elusive birds, this endemic lowland cuckoo is strictly terrestrial, although it will roost and nest in low trees. This large, almost pheasant-like cuckoo can be difficult to observe due to its wary and secretive habits and its preference for dense lowland swamp and alluvial forests. Its appearance and vocalisations are unmistakable. The sexes are alike with glossy green-and-purple upperparts and fine black-and-white barring on the belly. The iris is surrounded by an extensive patch of pale green skin, matching the colour of the bill. The song is a low-pitched, far-carrying tremulous *koo* or *wu-koo*, while the call is a pig-like grunt that can generally only be heard at close quarters.

**Where to see** Endemic. A rarely observed lowland specialist, exclusively found below 900m (and mainly <500m). Most records are from Sabah.

# Greater Coucal
*Centropus sinensis* 50cm

This black-and-chestnut bird could
be mistaken on Borneo for the closely
related Lesser Coucal, which is smaller,
often with pale streaks on the head
and wings, or for Short-toed Coucal,
which is smaller with a shorter tail and
different vocalisations. Greater Coucal is
most often seen in grasslands, thickets,
plantations and gardens, where it can be
observed clambering around in search of
small vertebrates and large invertebrates,
occasionally making short laboured
flights.

**Where to see** Common throughout, often
in pairs.

# Grey-rumped Treeswift  *Hemiprocne longipennis*  20cm

As the name suggests, treeswifts resemble
swifts, but they differ in several ways, the
most notable being that they are often
seen perched on prominent branches,
from where they sally out to catch insects
on the wing. In flight the grey rump
contrasts with the darker upperparts. The
male has dark red ear-coverts, while the
females are blackish.

**Where to see** Can be seen in small
flocks foraging over water or in clearings
throughout lowland areas of the island.

# Whiskered Treeswift *Hemiprocne comata* 16cm

This species could be mistaken for the previous species in flight but when perched it is unmistakable. The facial plumage is quite remarkable – the long white feathers of the supercilium and cheeks form a set of 'whiskers'. Another lowland species, its behaviour differs slightly from its congener in that it tends to fly at lower levels and for shorter periods of time.

**Where to see** All forested areas throughout the island.

# Edible-nest Swiftlet *Aerodromus fuciphagus* 12cm

This is the famous swiftlet that supplies the main ingredient in the bird nest soup trade, namely the saliva that solidifies to make the nests. It is a small blackish-brown swiftlet that is difficult to differentiate from the closely related Mossy and Black-nest Swiftlets, except when on the nest. Edible-nest Swiftlets form very large breeding colonies in cave systems and use echolocation to navigate and hunt insect prey.

**Where to see** They occur in a wide variety of habitats throughout the island. In recent years artificial bird houses have become common, broadcasting loud recordings of the birds' vocalisations in order to attract them for nest-farming.

Swifts

# Plume-toed Swiftlet
*Collocalia affinis*   10cm

A tiny swiftlet that is only seen in flight
except when on the nest, which is built
in areas with overhangs, including on
buildings. This species is more likely to
be seen flying at low levels than any other
swiftlet, and the glossy bluish plumage
with contrasting white belly is distinctive.

**Where to see** Common resident in all
habitat types on Borneo.

# Silver-rumped Spinetail
*Rhaphidura leucopygialis* 11cm

A small needletail with contrasting glossy black upperparts and a white uppertail. The shape in flight is distinctive and it is most often seen flying low over water in small flocks. These birds also forage at higher elevations, up to 40m over forested areas. This aerial insectivore flies rapidly, often with a side-to-side rocking motion.

**Where to see** A locally common resident throughout.

# Asian Palm Swift
*Cypsiurus balasiensis* 13cm

This swift is indeed closely associated with palm trees, where it nests and roosts. The relatively long wings and long, thin forked tail distinguish it from other swifts on the island. They are gregarious and often form very large flocks. An aerial insectivore, its high-pitched trilling calls are also notable.

**Where to see** Locally common in lowland areas where palms are found.

**Swifts**

# House Swift
*Apus nipalensis*  15cm

As the name suggests, this small, dark swift is closely associated with human habitation. The whitish throat patch, white rump and shallow forked tail – as well as habitat – distinguish it from other swift species. They are highly gregarious high-flying aerial insectivores, often forming huge colonies in towns and cities.

**Where to see** A common resident throughout.

# Large Frogmouth
*Batrachostomus auritus*  40cm

The largest of the frogmouths on Borneo, this is nevertheless a very secretive nocturnal bird relying on its incredibly cryptic (camouflaged) plumage to remain hidden during the day. The name 'frogmouth' derives from its wide frog-like mouth, which allows it to catch all types of prey, from insects to lizards and small mammals, by pouncing on them on the ground. This frogmouth is brownish with a paler nuchal collar and bold white spots on the wings and breast.

**Where to see** A locally rare lowland resident throughout.

# Sunda Frogmouth
*Batrachostomus cornutus* 25cm

This frogmouth occurs in two morphs – a rufous and a tawny-brown morph. The more common brown form shows a broad whitish supercilium and a white collar with black-and-white speckling on the back. The female has smaller white spots on the scapulars and fewer spots on the underparts. The call is a short, clear and high-pitched whistled *phyu*.

**Where to see** A rare lowland resident throughout.

# Gould's Frogmouth
*Batrachostomus stellatus* 23cm

Two colour morphs occur – a rufous form and a dark form. Both have a buffish supercilium and a narrow white collar. As with all frogmouths, the colour and patterns are cryptic, and these are secretive, nocturnal predators. They are most easily detected by their distinctive throaty two-note *wuuuu-waa* calls.

**Where to see** A locally scarce lowland resident throughout.

# Large-tailed Nightjar  *Caprimulgus macrurus*  27cm

The nightjars are a group of nocturnal birds that hunt insects at night and roost on the ground during the day, hidden with the aid of their cryptic plumage. They are most likely to be seen at dusk when they take to the wing but they look similar to other nightjars in the region so listen for the unusual and distinctive *chonk chonk* calls.

**Where to see** Found in all types of lowland habitat throughout the island; they can sometimes be seen on roads or tracks at night as they rest on the ground.

# White-breasted Waterhen
*Amaurornis phoenicurus*   30cm

This somewhat chicken-like bird is an inhabitant of all kinds of wetlands – from the margins of rivers and lakes to swamps, paddyfields and drainage ditches. It is distinctive, with blackish upperparts, white from the face to the belly and rufous under the tail. It is often seen darting from dense cover onto roadsides or trails. Its extraordinary array of calls includes loud grunts, croaks, bubbling gurgles and a *kuwaa kuwaa* territorial call.

**Where to see** A very common resident throughout.

# Ruddy-breasted Crake
*Porzana fusca*   22cm

A small, secretive rail with olive-brown upperparts and rufous face and underparts. The legs and feet are bright red, as are the eyes. It inhabits marshes, swamps, paddyfields and ponds where it furtively hunts for insects, molluscs and other small invertebrates under dense cover.

**Where to see** A rare lowland resident, or possibly a migrant, with an uncertain distribution on Borneo.

Rails, Gallinules and Coots

# Pacific Golden Plover *Pluvialis fulva* 24cm

A migratory shorebird rarely seen in breeding plumage on Borneo. In non-breeding plumage it is brown with gold speckling, paler on the underparts and face and with a buffish-white supercilium. In flight the underwings are pale grey. It is quite similar to the closely related but less widespread Grey Plover, which is larger and greyer and has black 'armpits' visible in flight.

**Where to see** A common visitor and passage migrant throughout from early September to April.

## Greater Sand-Plover
*Charadrius leschenaultii* 23cm

A large *Charadrius* plover with rather plain brown upperparts in non-breeding plumage and a robust bill and pale legs. The forehead and indistinct supercilium are white as are the underparts, with the breast-sides greyish-brown. In flight there is a white bar on the upperwing and white sides to the rump. Lesser Sand-Plover is smaller, with a more delicate bill and shorter legs.

**Where to see** Very common non-breeding visitor to coastal areas throughout, from September to April. A few may remain year-round.

# Eurasian Whimbrel *Numenius phaeopus* 43cm

This medium-sized curlew is notable for its long decurved bill and distinctive call. Whimbrel can be seen on mudflats, estuaries, mangroves and sandy beaches, where they glean and probe the mud or sand for crabs and other invertebrates with their superbly adapted bills. The voice is a highly distinctive tittering, musical *ti-ti-ti-ti-ti...*, usually given in flight.

**Where to see** Common non-breeding visitor throughout; dates are uncertain as some remain during the Northern Hemisphere summer.

## Common Greenshank
*Tringa nebularia* 32cm

This large wader has a thick, slightly upturned black bill with a yellowish-green base and greenish legs. In non-breeding plumage the head and back are sandy-grey, and the underparts are whitish. In flight the bird shows a long white area from the back to uppertail. Generally solitary, it forages for small aquatic invertebrates and even small fish in shallow water.

**Where to see** Common (Sabah, Brunei) to uncommon (Sarawak, Kalimantan) on coastal wetlands, beaches, grasslands and paddyfields from September to early June.

# Wood Sandpiper *Tringa glareola* 20cm

A medium-sized wader with sandy-grey plumage and paler underparts. The patterning on the back gives it a somewhat spotted appearance. A long white supercilium extends behind the eye, and the longer legs differentiate it from the similar Green Sandpiper. In flight the underwings are whitish, and the feet project well beyond tail. They feed by probing and picking for small aquatic invertebrates in shallow water.

**Where to see:** Very common non-breeding visitor and passage migrant to coastal and interior areas throughout, from August to April.

# Common Sandpiper *Actitis hypoleucos* 20cm

A dumpy, medium-sized wader with a distinctive tail-bobbing action. It has an indistinct long white supercilium, and white underparts with brownish patches on the breast-sides. This wader can be found in a wider range of habitats than others, including beaches, mangroves and paddyfields from sea level to 1,300m. It is often seen singly or in small loose groups running along muddy riverbanks in the interior.

**Where to see** Common non-breeding visitor throughout coastal and interior Borneo, from August to May; small numbers remain year-round.

# Grey-tailed Tattler   *Tringa brevipes*   25cm

A medium-large grey wader with a long pale grey supercilium and a dark grey line across the lores. It can be found on sandy beaches, estuaries, mudflats and mangroves, but especially favours rocky foreshores. Usually seen singly, it avoids freshwater habitats. Walks with bobbing and teetering movements.

**Where to see** Common non-breeding visitor and passage migrant to coastal northern Borneo (Sabah, Brunei, Sarawak, E. Kalimantan) from September to May; some remain year-round.

# Red-necked Stint   *Calidris ruficollis*   15cm

These tiny waders can often be seen in large numbers as they feed by probing very actively in mud and sand on mudflats, estuaries and sandy beaches. They are usually seen only in non-breeding plumage on Borneo: grey upperparts with broad darker streaking; white underparts with greyish patches on the breast-sides; and a white supercilium.

**Where to see** A common non-breeding visitor and passage migrant to coastal N. Borneo, from August to May.

# Crested Tern *Thalasseus bergii* 48cm

A large, stocky tern with a robust yellowish bill and dark grey upperparts. The black cap with shaggy crest is distinctive. In non-breeding plumage the forehead to crown is white, and the black crest shows whitish streaks. It flies rapidly but with deep, slow wingbeats, hunting small fish by plunge-diving. Often seen perched on driftwood and posts in shallow estuaries.

**Where to see** Common non-breeding visitor and occasional breeder throughout on open seas, coastal beaches, mudflats and estuaries.

## Black-naped Tern *Sterna sumatrana* 32cm

A small, dainty tern with a conspicuous black nape-band and a long, pointed tail, which in flight is deeply forked. The head is white with black bands from the eyes joining on the nape. In non-breeding plumage there are dark streaks on the hindcrown. They can be seen on rocky islets and beaches, but only rarely far out to sea. Often perch on lone offshore rocks, posts or buoys.

**Where to see** A locally common coastal resident throughout.

# Oriental Darter   *Anhinga melanogaster*   90cm

This cormorant-like bird has a long, snake-like neck and spear-like bill, as well as an elongated body and long tail, with a distinct kink in the neck. It is sometimes colloquially known as 'snake-bird'. It hunts by swimming underwater to spear fish with its sharp bill; usually only the head and neck are visible above the waterline. Perches on open branches with wings outstretched to dry.

**Where to see** Locally common to rare but declining resident in all types of wetlands.

**Storks**

# Storm's Stork *Ciconia stormi* 85cm

An unusual-looking, long-legged and secretive bird that feeds at small pools within forest. It can sometimes be seen perched on prominent branches in tall trees and often soars in groups on thermals over waterways. It has a unique face pattern with a bright red eye surrounded by broad bright yellow orbital ring, dull orange facial skin, and a long straight bright orange-red bill.

**Where to see** A rare and local resident in lowland wetlands. Borneo, especially Sabah, is a stronghold for this Endangered species.

# Great-billed Heron   *Ardea sumatrana*   110cm

Much larger than any other heron on the island, this uniformly brown-grey waterbird has a large bill, elongated back plumes and a long nuchal crest. It is usually solitary or in pairs and can be found on rivers, mangroves, estuaries and coastal mudflats. It often waits motionless for prey for long periods and gives a loud, harsh croak on taking flight.

**Where to see** Locally uncommon but widespread in Sabah; very uncommon in Brunei and Sarawak.

**Herons, Egrets and Bitterns**

## Lesser Adjutant
*Leptopitilos javanicus* 115cm

This large stork has a bulky, dirty yellow
bill, a bare pale pink head and yellow
neck with frilly feathers, and a white
neck-collar contrasting with the glossy
black back. It flies with its neck retracted
and long legs trailing behind. Often in
small groups around waterways, either
soaring on thermals or foraging on
the ground for small to medium-sized
invertebrates.

**Where to see** Uncommon and local
resident throughout the lowlands.

## Grey Heron  *Ardea cinerea*  95cm

A large, mainly grey bird, with a white
head and neck and prominent black lines
from eyes to long pendant crest, black
streaks on foreneck; the underparts are
white with black flanks. They are found
in estuaries, rivers, freshwater swamps
and ponds, and paddyfields. Usually
solitary. The flight is slow and laboured.

**Where to see** Mostly coastal on Borneo:
a regular but uncommon non-breeding
visitor to Sabah and Sarawak from
November to mid-April; rare in Brunei
and E. Kalimantan.

# Eastern Great Egret
*Ardea alba*  90cm

A large all-white heron with a heavy bill, a long gape-line extending behind rear of eye and a long, kinked neck. Shows a yellow iris, orange-yellow facial skin sometimes tinged blue, bright yellow bill and black legs. In breeding plumage, it develops long back plumes, a black bill, deep greenish-blue facial skin, and pinkish-red legs. It is found in all wetland habitats.

**Where to see** Common non-breeding visitor and possibly rare resident throughout, less common on the east coast.

# Little Egret
*Egretta garzetta*  60cm

A small, delicate all-white heron with a long slender grey-black bill and greenish-yellow facial skin. In the breeding season it develops two long white nape plumes, and filamentous dorsal plumes extending to the end of tail. Two subspecies occur on Borneo, one with all-black legs (nigripes) and feet, and one with black legs and yellow feet. It is an active feeder in shallow wetlands, mangroves, lakes, rivers, swamps and paddyfields.

**Where to see** Uncommon resident and common non-breeding visitor throughout; both subspecies probably occur in roughly equal numbers.

# Pacific Reef Heron *Egretta sacra* 60cm

This heron is unusual in that the adults occur in two different forms, which occur in roughly equal numbers. The white morph is pure white, while the dark morph is a dark slaty bluish-grey to brownish-grey. The bill is blackish-brown with a yellowish base and the legs are pale yellowish-green to greenish-black. The white morph can be confused with Little Egret, which has a black bill and yellow feet (except *nigripes*), or Eastern Cattle Egret, which is smaller with a shorter bill.

**Where to see** Almost exclusively coastal.

# Eastern Cattle Egret *Ardea ibis* 50cm

An all-white heron, sometimes with some inconspicuous buff on the crown, no plumes and a short thick neck with bulging throat; the bill is yellow with a distinctive blunt shape. This species doesn't breed in Borneo but may sometimes be seen in breeding plumage, where the head, neck, breast and back are pinkish-buff, sometimes with pinkish-orange plumes on the head. This species is usually seen in flocks in fields and pastures, associating with grazing livestock.

**Where to see** Common non-breeding visitor and passage migrant throughout.

# Striated Heron  *Butorides striata*  40cm

A small, hunched heron that is often flushed from vegetation or seen perched motionless at the water's edge. It is solitary and when flushed usually flies low. The male is slate-grey with a blackish crown and long black nape plumes. The female is like the male but has a brownish wash on the sides of the neck and flanks. Immature birds look quite different, being brownish with pale scaling and spots on the upperparts.

**Where to see** A common resident in wetlands throughout.

# Black-crowned Night-Heron *Nycticorax nycticorax* 60cm

This chunky heron can be seen roosting in loose colonies in trees near estuaries, mangroves, rivers, swamps and pastures during the day, flying out at dusk to feed in shallow wetlands. The crown and back are black with one to three long white nape plumes, grey wings and tail; underparts are paler. Immature birds are brown, with bold buff spots and streaks on the upperparts, and streaking on the breast.

**Where to see** Locally fairly common resident throughout.

# Cinnamon Bittern *Ixobrychus cinnamomeus* 40cm

Rather similar to Yellow Bittern but the upperparts are unmarked uniform rich cinnamon. In flight the back and wings are a uniform cinnamon colour. The female is duller with some white streaks on less rich cinnamon upperparts, and brown-streaked dull cinnamon underparts. Immatures are browner, barred and spotted with buff, and streaked below.

**Where to see** Locally common resident throughout in freshwater swamps and ponds, reedbeds and paddyfields.

# Yellow Bittern *Ixobrychus sinensis* 37cm

A small buffish-yellow bittern with a blackish crown and olive-brown upperparts. In flight the black flight feathers and tail contrast strongly with yellow-brown wing-coverts and dark brown back; underwing-coverts are white. Females and immatures are like the male but have a brownish crown and a more strongly streaked neck and back. This shy bird is most often seen flying short distances low over suitable habitat.

**Where to see** An uncommon resident and non-breeding visitor at margins of freshwater wetlands.

# Oriental Honey-buzzard
*Pernis ptilorhynchus* 60cm

This raptor's plumage is highly variable, but all forms have a longish tail and distinctive pigeon-like head on a long neck. The adults have a short crest on the hindcrown, and an off-white throat with dark gular stripe. In flight from below the wings are broad, barred rufous on the underside. Soars on flat wings, often twisting the head from side to side.

**Where to see** Common throughout in riparian forests, lowland forests, logged forest and plantations.

# Jerdon's Baza
*Aviceda jerdoni*  45cm

This medium-sized raptor has a distinctive long, black erect crest tipped white, which is usually held vertically. The head and neck-sides are rufous with dark streaking, while the belly and thighs have broad rich rufous bars. This species can be mistaken for Wallace's Hawk-Eagle, but the baza has less feathering on its legs and longer wings relative to the tail.

**Where to see** This uncommon resident is usually found in riparian and lowland forests.

# Bat Hawk
*Macheiramphus alcinus*  45cm

As the name suggests this unusual falcon-like raptor preys mainly on bats, which it catches and devours in flight. Bat Hawks are mainly crepuscular but are also known to hunt at night, especially when moonlit. It is an all-dark hawk except for a white bib on the throat and breast. Requires tall, bare-boled trees for nesting in high canopy, and an abundant source of bats or swiftlets.

**Where to see** Uncommon resident throughout in lowland, riparian and secondary forest; often associated with limestone outcrops and caves.

# Black-shouldered Kite
*Elanus caeruleus* 30cm

A small, graceful kite with immaculate plumage that can be seen in open country, paddyfields and pastures. It is mostly white with a grey back and wings, and a black shoulder-patch. Active throughout the day, but mostly crepuscular; often seen hovering over fields and open areas, feeding on rodents, grasshoppers, skinks and quails. The flight is graceful, with the wings raised when gliding.

**Where to see** Locally uncommon, possibly increasing, and resident throughout.

# Brahminy Kite *Haliastur indus* 48cm

With its chestnut plumage and contrasting white head, neck and breast, this is one of the easiest raptors to identify on Borneo. Immature birds can be a little confusing though, due to their indistinctly streaked brown plumage. This species is commonly seen in open areas and forest edges near estuaries, lakes, rivers and along coasts, where they feed on fish, snakes, lizards, large insects and carrion.

**Where to see** Common resident throughout; occurs inland but most abundant on coast.

# White-bellied Sea-Eagle
*Haliaeetus leucogaster*   80cm

A large, white, unmistakable bird of prey with a grey back and wings and black flight feathers. In flight the white underwing-coverts contrast with greyish-black flight feathers, and the wedge-shaped tail is distinctive. Immature birds are streaky brown but can be told from other raptors by their large size and tail shape. This opportunistic carnivore feeds on carrion as well as fish, birds and snakes, and has been known to steal prey from other raptors.

**Where to see** Common resident along entire coast, but less common on inland waterways; never far from water.

# Lesser Fish-Eagle
*Ichthyophaga humilis*   60cm

Greyish-brown with a grey head and white belly and thighs. The tail is brownish with a darker terminal band fading to a paler base on the undertail. It inhabits narrow, undisturbed forest watercourses, mangroves and plantations in the lowlands, perching in large trees overhanging narrow rivers, often flying off ahead of canoes travelling along rivers. Feeds on fish.

**Where to see** An uncommon resident throughout the interior.

# Grey-headed Fish-Eagle *Ichyophaga ichthyaetus* 70cm

A large, greyish-brown eagle with a diagnostic white tail and sharply demarcated broad black terminal band. They are found on rivers, lakes, lowland wetlands and occasionally on coasts and offshore islands. Tends to favour larger watercourses than the similar Lesser Fish-Eagle, which is also smaller with an all-greyish-brown tail. Grey-headed Fish-Eagles spend much time perched over water, swooping down to take fish or carrion from surface.

**Where to see** Uncommon resident throughout the lowlands.

# Mountain Serpent-Eagle
*Spilornis kinabaluensis* 55cm

Similar to Crested Serpent Eagle but larger, with darker plumage and a black throat. In flight the bands on the tail are bolder and broader, and the underwing-coverts are darker. This bird is found strictly in montane forests. Behaviour is similar to Crested Serpent Eagle. It is often seen soaring and circling over the forest canopy but hugs ridges. Feeds on snakes and lizards.

**Where to see** Endemic, rare resident in the north and north-central mountain ranges from Mount Kinabalu to Mount Murud and Mount Mulu.

# Black Eagle
*Ictinaetus malaiensis* 75cm

A large, all-black eagle with a yellow cere, and yellow feet. In flight the broad wings slightly narrow inwards, the long narrow tail shows inconspicuous darker banding and the yellow feet are conspicuous. They are very often seen in pairs, usually soaring over forested hillsides. Possibly specialises on predating canopy-dwelling birds, small mammals and reptiles.

**Where to see** Uncommon resident throughout primary lowland and hill forest.

# Crested Serpent-Eagle
*Spilornis cheela* 65cm

A brown raptor with a fan-shaped, short, rounded, black floppy crest at the back of the head. The plumage of the underparts is dotted with small white spots, and the yellow iris and bare yellow facial skin are distinctive. Its broad rounded wings are held slightly forward in flight. Feeds on snakes and small vertebrates. A noisy raptor, often giving a plaintive *kek kek kweee kweee kweee* call, especially in flight.

**Where to see** Common lowland resident throughout.

# Crested Goshawk
*Accipiter trivirgatus* 40–46cm

Despite its name, the short crest is often not visible. This large *Accipiter* is a high-speed hunter of small mammals, small birds, lizards and frogs, usually taking off in pursuit from a concealed perch. It has a distinctive display flight involving exaggerated stiff wing-flapping with wings depressed below the horizontal, alternating with normal flight. The back is brown while the underparts are barred and streaked with rufous and white. The female is larger and browner.

**Where to see** Common resident throughout in lowland to upper montane forest.

Imm

Juv

# Rufous-bellied Eagle  *Lophotriorchis kienerii*  60cm

A medium-sized eagle with a prominent black hood contrasting with the white throat. The black upperparts contrast with the white upper breast and deep rufous belly. Immature plumage looks quite different with a black mask, brown upperparts with buffy fringes, and all-white underparts. An aerial hunter that makes spectacular stoops into the forest, this species seems to favour forest edges and clearings.

**Where to see** Uncommon resident in primary and secondary forests, including plantations.

# Changeable Hawk-Eagle
*Nisaetus cirrhatus*  60cm

A striking, short-crested eagle with many different forms. As in many raptors, the female is larger than the male. The pale morph has dark brown upperparts with buff feather fringes, and the white underparts are boldly streaked dark brown. The dark morph is all blackish-brown and can be confused with the larger Black Eagle, which has all-dark underwings. The immature is paler with mottled brown upperparts.

**Where to see** Uncommon resident in lowland and hill forests throughout.

# Blyth's Hawk-Eagle
*Nisaetus alboniger*   52–58cm

Striking black-and-white plumage, with long, erect black crest tipped white, breast boldly streaked black and white, and belly barred black and white. In flight the underwings are white dotted black, and the white flight feathers show narrow black barring. The undertail is black with a broad white band. Immature birds have dark brown upperparts with buffish feather fringes, and plain pinkish-buff underparts becoming streaked later; almost identical to immature Wallace's Hawk-Eagle.

**Where to see** Uncommon resident in hill forest in N. Borneo.

# Wallace's Hawk-Eagle
*Nisaetus nanus*   43–58cm

The adult is very like subadult Blyth's Hawk-Eagle but distinguished by its smaller size and tail pattern, which has two pale and three dark bands. The female is larger than the male. The immature is almost indistinguishable from immature Blyth's Hawk-Eagle but is smaller with a broader white tip to the crest. A canopy dweller in lowland, riverine and swamp forests, as well as plantations.

**Where to see** An uncommon to locally common resident but appears to be absent from N. and W. Sabah.

# Oriental Bay-Owl *Phodilus badius* 35cm

An unmistakable owl with a squarish, pale pinkish-buff facial disc and ear-like projections above the eyes. The back and wings are chestnut with black and buff spots, and the pinkish-buff underparts are spotted dark brown and buff. Feeds on small vertebrates and insects, which it hunts from a perch in dense vegetation, rocking its head from side to side as it watches potential prey. The call is an ethereal, piercing *hwee hwee...*

**Where to see** Locally uncommon lowland resident throughout.

# Reddish Scops-Owl
*Otus rufescens*  16cm

The scops-owls are small owls with prominent ear-tufts, an indistinct facial disc fringed dark brown and cryptic plumage. All are nocturnal and very shy. The plumage of this species is rufous-brown with small buff spots and paler underparts. It is more spotted than the otherwise similar Mountain Scops-Owl and more rufous and spotted than Sunda Scops-Owl. All are best distinguished by their song; Reddish Scops-Owl has a single-noted, soft *hyoo*.

**Where to see** Uncommon resident in lowland and montane forest throughout.

# Mountain Scops-Owl
*Otus spilocephalus*  19cm

As the name suggests, this owl is confined to montane forest. Strictly nocturnal and very elusive; feeds on insects, small birds and rodents in the lower and midstorey of dense forest; roosts in dense vegetation. Song is a pure, even-pitched *plew-plew* whistle uttered every four or five seconds with a short interval between each note.

**Where to see** Locally common in north-central mountain ranges.

# Barred Eagle-Owl
*Bubo sumatranus*  44cm

A large owl with striking long brown ear-tufts, and finely barred underparts. Found in lowland forests, plantations and gardens. Occupies large territory and roosts in tall trees with dense foliage, usually close to trunk. Nests in tree hollows and is very site-faithful, returning to same nest year after year. The call is a series of hoots, also an eerie *whaa*.

**Where to see** Sparsely distributed and uncommon resident throughout.

# Buffy Fish-Owl
*Ketupa ketupu*  40cm

A dark brown owl with long ear-tufts held horizontal. Frequents forested areas near water including tree-lined rivers and lakes, estuaries, paddyfields, parks and mangroves. Hunts at night from perches near water, feeding on crustaceans, aquatic insects and small vertebrates. Roosts during the day in dense foliage in trees and palms. A variety of calls including a harsh, upslurred scream and a fluty, melancholy *pop pop....*

**Where to see** Locally common resident in lowlands throughout.

# Brown Wood-Owl
*Strix leptogrammica*   50cm

Easily distinguished by the rufous facial disc with broad, dark-brown-edged rings around eyes. The mantle and back are chestnut with dark brown barring, and the underparts are finely barred buffish-white. Strictly nocturnal in dense primary lowland and hill forests. Nests in tree hollows and roosts in the densest parts of tall trees. The call is a deep, mellow *boo-booo*.

**Where to see** Uncommon resident throughout.

# Brown Boobook
*Ninox scutulata*   30cm

A dark brown owl with buffish-white underparts with dark brown streaks. It is a lowland specialist in primary lowland forest and mangroves. Feeds on large insects and small vertebrates, often catching prey on the wing. Roosts during the day under thick canopy or among creepers; nests in tree hollows. The call is a distinctive, pleasant, mellow *hoo-wup*.

**Where to see** Common to uncommon resident throughout.

# Whitehead's Trogon
*Harpactes whiteheadi* 32cm

The only trogon found at high altitude
on Borneo. Very shy and unobtrusive,
and easily overlooked. It rarely vocalises.
Sallies out from a perch in the mid- to
upper storey below the canopy in dense,
damp forest for large insects, which
it gleans from foliage. Male has a red
crown. Distinctive underparts, with black
throat grading into grey breast clearly
demarcated from crimson below. The
female is duller, with head and lower
underparts cinnamon-brown.

**Where to see** Endemic, uncommon
resident in north-central mountain ranges
of N. Borneo.

♀

♂

♂

# Red-naped Trogon
*Harpactes kasumba* 32cm

A bright red bird with a conspicuous bright red nape-stripe that separates the black head from the golden-brown back. The reddish-brown iris is surrounded by broad blue orbital skin. The female is duller, with a greyish-brown head, neck and breast. Quiet and unobtrusive. Sallies out from a perch in the midstorey for large insects, which it takes on the wing. The song is a slow, sad-sounding 5–8-note series, *pau pau pau pau pau...*

**Where to see** Common resident throughout in lowland forest.

# Diard's Trogon
*Harpactes diardii* 34cm

Similar to Red-naped Trogon but has a dark purple hindcrown, pink nape-stripe, pinkish breast-band, and violet skin around the eye. The female has an olive-brown head, breast and upperparts and pinkish underparts. Its behaviour is similar to Red-naped Trogon. The song is a mournful, quite rapid series of 15–20 *pau* notes, lower-pitched and faster than that of Red-naped Trogon. Found in lowland, riparian to lower montane forest.

**Where to see** Common resident throughout.

# Scarlet-rumped Trogon
*Harpactes duvaucelii*  25cm

A small trogon with scarlet underparts contrasting with cinnamon-brown upperparts. The male has an entirely black head and scarlet rump as well as a bright blue bill, gape and skin above eye. The female is plain olive-brown with pinkish underparts and less blue on the face. The distinctive song is a rapid, 'bouncing ping-pong ball' song, a series of 18–20 short, rapid notes that descend in pitch.

**Where to see** Common lowland resident throughout.

♀

♂

# Oriental Pied Hornbill
*Anthracoceros albirostris*  60cm

A pied hornbill with glossy black upperparts and breast and bluish-white facial skin and throat patch. The large bill is pale yellow with a black base to lower mandible. The female is smaller with a less pronounced casque. Feeds mainly on fruit, also insects, spiders, small vertebrates. A variable, harsh, loud yelping *kek kek kuck kuck,* reminiscent of a small barking dog.

**Where to see** A common lowland specialist in secondary, peat swamp, riverine and coastal forests, and plantations.

# Black Hornbill
*Anthracoceros malayanus*   65cm

Black, sometimes with a broad white to grey stripe over eye to nape, a long tail with broad white tips to outer tail feathers and a pale yellow bill and casque. In flight it is all black with a small patch of white near tail tip. A lowland specialist found in primary and secondary forest, coastal forests, and plantations; tolerant of degraded forest. Usually seen in pairs. Mainly feeds on large fruit, but also insects and small vertebrates.

**Where to see** Locally common resident.

# Rhinoceros Hornbill
*Buceros rhinoceros*   85cm

A spectacular, large hornbill with an impressive orange-and-red bill and casque. The casque is cylindrical with a narrow black stripe on each side and a strongly recurved tip, with considerable individual variation. The plumage is black with a white belly and thighs. The white tail has a broad black band across the centre. The male has a red iris, whereas the female's is white with a red eye-ring. Usually in pairs. Often starts calling just before taking flight; utters a loud, deep, far-carrying bark *grark*.

**Where to see** Locally common resident in suitable habitat throughout.

# Helmeted Hornbill

*Rhinoplax vigil* 110–120cm

An unmistakable, huge hornbill with an exceptionally long tail and unique heavy, blunt, yellow casque with red sides and base and bare maroon-red wrinkly skin around the face and neck. It is the only hornbill with a solid ivory casque. The female is smaller with a bare, wrinkly turquoise throat. Usually in pairs. Feeds mainly on fruit, especially figs, but also forages in the canopy of tall trees for small animals, including squirrels, birds and lizards. Its unique, very far-carrying vocalisations are given from the canopy and immediately recognisable.

**Where to see** Uncommon resident throughout in primary lowland and hill forest (avoids secondary forest).

# Bushy-crested Hornbill
*Anorrhinus galeritus*   60cm

The only hornbill with no white in the plumage. Blackish-brown with long, loose crown and nape feathers forming a droopy crest, and bare pale blue skin around the eye and throat. The bill and low, inconspicuous casque are black. The female is smaller, with a yellow bill and casque. Unlike other hornbills on Borneo, they are usually encountered in noisy groups, sometimes of up to 20 birds.

**Where to see** Common resident throughout in lowland forest, favouring foothills with high density of fig trees.

# Wrinkled Hornbill *Rhabdotorrhinus corrugatus* 70cm

Black with a white face, breast and tail, which are often stained ochre. Bare blue skin around eye, yellow gular pouch and yellow bill with a red, ridged base and wrinkled casque. The female is smaller with a black head and neck and blue gular pouch, as well as a smaller bill and reduced all-yellow casque. Favours swamp forest; highly mobile and nomadic, tracking fruiting trees, usually in pairs.

**Where to see** Uncommon resident throughout; scarcer on west coast.

# White-crowned Hornbill *Berenicornus cornatus* 75cm

Another spectacular hornbill with a distinctive shaggy, erect white crest and blue facial skin. The male is white with a black back and wings. The female is smaller with a reduced crest and black underparts. Inconspicuous, usually remaining well below canopy and regularly descending to ground; territorial, usually in groups of 4–8. More often heard than seen due to its habit of remaining below canopy; the call is an owl- or pigeon-like mellow, low-pitched, three-note *hu hoo-hoo*.

**Where to see** Uncommon to rare resident throughout at low densities.

# Wreathed Hornbill
*Aceros undulatus*  80cm

Black with a white face to upper breast and short white tail. Red skin around the eye; the bill and casque are pale yellow with ridged corrugations; the gular pouch is yellow. The female is smaller, with a black head and neck, and blue gular pouch. Requires large, unbroken tracts of forest. Flies high above the forest canopy, rarely gliding as other large hornbills do. A raucous, loud, grunting *uk-guk*. The very loud, whooshing wingbeats are distinctive.

**Where to see** Nomadic, locally common resident throughout; patchy distribution due to forest loss.

# Gold-whiskered Barbet
*Psilopogon chrysopogon*  30cm

The largest barbet on Borneo is easily distinguished from other similar-looking barbets by the large yellow patch on the side of the face. The head has a complex pattern with a yellow forehead, red crown and black-and-blue mask around the eyes. The rest of the body is bright green. As in all barbets, the bill is large with black bristles around the base. It forages in the canopy for figs and other fruits in the lowlands. Calls are introduced by a rapid *tu-tu-tu-tu-tu...*, usually followed by a resonant, mellow *tuu-tu-tu-toop*, often repeated endlessly.

**Where to see** Common resident throughout.

# Blue-eared Barbet
*Psilopogon duvaucelii* 16cm

This and the similar Bornean Barbet are the smallest of the barbets on the island. Unfortunately, the name is confusing as the subspecies on Borneo has black ear-coverts rather than blue. To make matters even worse, the Bornean Barbet has blue ear-coverts! Blue-eared Barbet also has a blue throat and crown (black and red respectively in Bornean). The call is a rapid, incessant two-note *tu-tuk tu-tuk tu-tuk tu-tuk*.

**Where to see** Common in lowland and hill forests, as well as mangroves and plantations.

# Mountain Barbet  *Psilopogon monticola*  22cm

As the name suggests, this barbet is found at higher altitudes. It is the least brightly coloured of the Bornean barbets, with a characteristic streaky forehead and throat. With practice, the barbets can readily be identified by their vocalisations. This species gives a rapid, resonant, repetitive *tu-ruk tuk tuk tuk tuk tuk tuk tuk tuk...* uttered incessantly.

**Where to see** Endemic, locally common resident above 750m throughout.

# Brown Barbet  *Caloramphus fuliginosus*  18cm

This small, brown barbet is gregarious, usually being found in vocal groups of 6–8 birds. The upperparts are dark brown, the throat and breast pinkish-red fading into dirty buffish-brown underparts. The bright orange-red legs and feet are conspicuous. The bill is long and slightly downcurved. The female's bill is much paler than the male's. The call is a thin, sibilant, rather soft but high-pitched *tseet tseet tseet*.

**Where to see** Endemic, common resident in lowland forests throughout.

# Golden-naped Barbet  *Psilopogon pulcherrimus*  21cm

The only barbet on Borneo with a plain blue crown and throat. The name comes from the golden-yellow stripe around the back of the neck. The upperparts are deep green, the underparts paler green. A montane barbet, usually found above 1,100m. Forages in the midstorey and canopy for fruits and occasionally takes insects. The call is a relatively high-pitched, mellow three-note *tuk tuk tukrrrk*.

**Where to see** Endemic, common resident in north-central montane ranges.

## Rufous Piculet
*Sasia abnormis*   9cm

A tiny rufous-coloured woodpecker with olive-green upperparts. The male has a yellow forehead, which in the female is rufous. Usually seen singly or in pairs; joins mixed feeding flocks; forages inconspicuously in dense undergrowth where it feeds on insects, gleaning prey from bark, bamboo, rattan and dead wood. It is often detected by its loud and persistent tapping. The voice is a squeaky, high-pitched one-note *peet* or two-note *pee-peet* (like child's toy).

**Where to see** Common resident in lowland and hill forests.

## Grey-capped Pygmy Woodpecker   *Dendrocopus canicapillis*   15cm

A small woodpecker with a black-and-white barred back; the crown is dark grey with a red spot at each side. The female lacks red on the crown. The similar Sunda Pygmy Woodpecker is greyer and duller overall. Forages in small branches of large trees and bushes, gleaning insects from bark, lichen and leaves, and hammering dead branches.

**Where to see** Common resident throughout in lowland forests and plantations; favours inland forest to 400m.

# White-bellied Woodpecker · *Dryocopus javensis* · 45cm

A very large black-and-white woodpecker with a crimson crest and broad stripe on the face. The female has less crimson on the head, it being confined to crest. Noisy and conspicuous; usually in pairs. Forages at all levels from upper limbs and trunks of tall trees to fallen timber on ground, nearly always on dead wood. The vocalisations include a short, loud far-carrying yelp *kiau* and a rapid *ke-ke-ke-ke*....

**Where to see** Fairly common lowland specialist resident in forest throughout.

## Rufous Woodpecker  *Micropternus brachyurus*  25cm

Rufous with narrow black bars on wings, rump and flanks, and a crimson patch below the eye. The female lacks any red on head. Often in pairs; very active with bounding and dipping flight. Forages at all levels from ground to upper canopy, feeding mainly on ants and termites. Very vocal; utters a drawn-out series of 3–5 screaming notes *kee-kee-kee-kee-kee...* with a plaintive quality.

**Where to see** Common resident throughout in lowland forests.

♂

## Banded Yellownape  *Chrysophlegma mineaceum*  25cm

♀

Olive-brown with buff scaling and red wing-coverts, and a yellow crest at the back of the head. The female is browner on the head, with buff spots on the cheeks. Feeds predominantly on ants and ant larvae, and actively forages in dense vegetation where it pecks, gleans and probes in vines, epiphytes, dead wood and on branches. Call a short, explosive *kau*.

**Where to see** Common resident throughout in lowland forests, mangroves, plantations and gardens.

# Checker-throated Woodpecker *Chrysophlegma mentale* 27cm

A dark green woodpecker with deep red wing-coverts and a yellow crest at the back of the head. It differs from the similar Crimson-winged Woodpecker in having a brownish submoustachial stripe which joins a broad chestnut line from the neck-sides to the breast. The black-and-white dappled throat gives the bird its name. It occurs in lowland to upper montane forests. The three- or four-note call *ki kee kee kee* has a somewhat sad, pleading tone.

**Where to see** Locally common resident throughout.

# Crimson-winged Woodpecker *Picus puniceus* 25cm

Olive-green with crimson wings and crown and a yellow crest at the back of the head. The male has a short crimson stripe on each side of the face and buff bars on the flanks. The female lacks the face stripe. Mostly in pairs; joins mixed feeding flocks; eats mainly ants and termites, including larvae. The song is a laughing cackle, *kii kii kii kii ki*, descending in pitch and volume.

**Where to see** Uncommon resident throughout in lowland to upper montane forests.

# Maroon Woodpecker
*Blythipicus rubiginosus* 23cm

The male is olive-brown with a purplish back and wings, and red sides of the neck and nape; long yellow bill. The female has a shorter bill and no red on the head. Found singly or in pairs; it feeds mainly on insect larvae and ants, foraging in the lower and midstoreys. The call is a short sharp squeaky *kik* repeated rather slowly, higher in pitch than other similar-sized woodpeckers.

**Where to see** Common resident throughout in lowland to upper montane forests, mangroves, plantations and gardens.

# Buff-rumped Woodpecker
*Meiglyptes tristis* 18cm

A small woodpecker with narrow buff and black barring over the entire body, broader bars on the mantle and wings and a crimson stripe on the sides of the face (absent on female). The buff rump is not always visible. Often in mixed feeding flocks; it feeds mainly on ants and other insects in small branches of the upper canopy, favouring forest edges, clearings and gaps.

**Where to see** Common resident throughout in lowland to hill forests.

# Orange-backed Woodpecker  *Reinwardtipicus validus*  30cm

The males and females of this striking woodpecker differ markedly. The male has a dark red crown and crest, and is white from the nape to the back with an orange wash on the lower back blending into the orange rump. The wings are blackish-brown with three to five broad reddish-brown bars; underparts dark red. The female has a dark brown crest, white rump and greyish-brown underparts. Nearly always in pairs, it is a dead-wood specialist foraging mostly in the midstorey. The voice is a series of squeaky, clucking notes *chik-chik-chik-chik-kichik*; drums in soft, very short bursts.

**Where to see** Common in the north, uncommon elsewhere in lowland to upper montane forests.

# Buff-necked Woodpecker  *Meiglyptes tukki*  21cm

Like the previous species, this woodpecker has narrow buff and brown bars over the entire body but differs in having a plain brown head. The male has a red malar stripe on each side of the face and a large buff patch on the neck-sides. The female lacks the red malar stripe. Could be mistaken for Buff-rumped Woodpecker but is larger, has prominent buffish neck-patch and lacks the plain buffish-brown rump.

**Where to see** Common resident throughout in lowland forests and plantations.

# Great Slaty Woodpecker  *Mulleripicus pulverulentus*  50cm

The world's largest woodpecker. A spectacular, prehistoric-looking, slaty-grey bird with a buffish-yellow chin and throat, and prominent red cheek patch. The female lacks the red cheek patch. Encountered in pairs or family groups of three to seven birds; they fly noisily from tree to tree and will often flare their wings as they climb in jumps up a prominent tree trunk. Feeds on ants and wood-boring larvae mainly in upper storey; forages in emergent trees, requiring large live trees for foraging on stingless bee, ant and termite nests. The voice is a loud, far-carrying, whinnying *wit-wit-wit-wee*.

**Where to see** Locally uncommon lowland specialist intolerant of logged forest.

# Blue-eared Kingfisher
*Alcedo meninting*   16cm

A small, jewel-like ultramarine bird with contrasting brilliant turquoise-blue back and uppertail, white neck-stripe and throat, and rich rufous underparts. Solitary; hunts from a perch over water from which it dives steeply to catch aquatic invertebrates and small fish with its sharp bill. Usually first detected by the short, high-pitched *tseet* call uttered in flight.

**Where to see** Common to uncommon resident throughout along streams and creeks, mainly in the lowlands.

# Oriental Dwarf-Kingfisher  *Ceyx eritheca*  14cm

A tiny kingfisher with all-rufous wings and bright orange-yellow underparts. The subspecies in Sabah has blue-black wings. The crown to uppertail is lilac-rufous. Solitary; tiny and fast-moving, it predominantly feeds on insects, but also aquatic invertebrates, small fish and frogs. It nests in excavated burrows in steep banks, termite mounds or fallen trees, not always close to water.

**Where to see** Common resident throughout in lowland to lower montane forest, often in lower to midstorey of dense forest far from water.

Kingfishers

# Banded Kingfisher  *Lacedo pulchella*  24cm

Males and females of this striking
kingfisher differ markedly. The male
is banded silvery-blue and black, with
bright orange-rufous underparts and a
shaggy blue crown. The female is banded
rufous and black, with buffish-white
underparts and narrow black bars on the
breast and flanks. Both sport a heavy,
bright red bill. It favours dense forest
away from water where it feeds on large
invertebrates, small fish and lizards. More
often heard than seen. The song is a long,
mournful series of *tu-wee* notes, the first
part longest and loudest, speeding up and
tapering off in volume towards the end.

**Where to see** Uncommon to rare
resident throughout in lowland and lower
montane forests.

# Stork-billed Kingfisher *Pelargopsis capensis* 35cm

A large and unmistakable kingfisher with a huge red bill. The head and underparts are buffish-orange, and the wings are greenish-blue. In flight the bright azure-blue lower back is eyecatching. Solitary; feeds on fish and crustaceans, hunting from a concealed perch over water. The song is a piercing but plaintive, two-note whistle *fyu fyu*.

**Where to see** Common lowland resident throughout favouring forested waterways, mangroves and wooded seashores.

# Collared Kingfisher
*Todiramphus chloris*  25cm

This very vocal kingfisher can often be heard giving its coarse, grating *ke kek kek kek kek kek* call. The bird is greenish-blue with a white collar joining white underparts. Noisy and conspicuous, defending its territory year-round, it feeds on crustaceans, insects, small fish, frogs and lizards.

**Where to see:** Common to abundant resident throughout in open areas in coastal woodlands, mangroves, paddyfields, plantations and gardens.

# Blue-throated Bee-eater
*Merops viridis*  22cm

A medium-sized green bee-eater with a prominent dark chestnut cap and long central tail-streamers. In flight the wings are long and pointed and the underwings are pale rufous with a blackish trailing edge. Often in flocks, they feed on flying insects, especially bees, wasps and dragonflies, hawking their prey from prominent perches. The voice is a pleasant, liquid *prrp prrp...*

**Where to see** Common in open areas in woodlands, marshes, mangroves, paddyfields, pastures, plantations and gardens in the lowlands.

# Red-bearded Bee-eater
*Nyctyornis amictus* 30cm

An extraordinary, bright green forest-dwelling bee-eater with a heavy decurved bill and a brilliant crimson 'beard'. The body is emerald-green, with a bright orange-yellow undertail with a black tip. The forehead and crown are pinkish-lilac (the female has a red forehead and smaller lilac crown patch), making this a very colourful bird! It can be seen hawking arboreal insects from high leafy perches in the midstorey near forest gaps. The call is a very unusual, throaty and gruff *kwek-kwek-kwek-kwek-kwek-kwe-kwe-kwk*.

**Where to see** Locally common to uncommon resident throughout in lowland to lower montane forests.

♀

# Oriental Dollarbird
*Eurystomus orientalis* 30cm

A stocky, large green-blue roller with a broad, short red bill and conspicuous white circles in wings. The name comes from these large, round white patches on the long, broad wings, which can be seen in flight. It favours high perches in dead trees from where it feeds on large insects caught on the wing. The voice is a short, raspy, repeated *chak-chak-chak....*

**Where to see** Common on forested riverbanks, edges of primary and secondary lowland forest, mangroves and plantations.

# White-fronted Falconet
*Microhierax latifrons* 16cm

A tiny black-and-white raptor with a prominent white forehead and buffish wash on the underparts. The sexes are similar but the female has chestnut forehead. Feeds on large insects, small birds and lizards, hunting from a high, exposed perch, sallying to take prey in air and often returning to the same perch for prolonged periods. Frequently bobs head when perched. Often in pairs or small groups. Flies with rapid wingbeats.

**Where to see** Endemic, uncommon resident restricted to N. Borneo.

♂

♀

# Peregrine Falcon
*Falco peregrinus*  34–48cm

A compact, robust falcon with long pointed wings. The resident race has a black head with solid black sides giving a hooded appearance. The larger migratory race has a black head with black moustachial stripes. Female is larger than the male. Feeds mainly on small birds and bats; characteristically very agile, yet strong, direct flight. Nests on cliffs.

**Where to see** The resident race is a rare, mostly montane resident in N Borneo; the migratory race is a non-breeding visitor throughout from October to April.

## Long-tailed Parakeet  *Psittacula longicauda*  45cm

A long-tailed green parrot with rose-pink cheeks to nape and a black throat. The elongated central tail is blue with green outer feathers. The male has a bicoloured red-and-black bill. The female has duller pink cheeks, a green nape and an all-black bill. Gregarious, usually in small flocks but sometimes in large congregations where there are fruiting trees; feeds on fruit, seeds and flowers.

**Where to see** A locally common extreme lowland specialist to 200m, found in riverine forests, mangroves, plantations and gardens; shuns primary forest.

♀

# Blue-crowned Hanging Parrot *Loriculus galgulus* 12cm

A tiny, bright green parrot with a dark blue spot on the crown, triangular golden patch on the mantle and a yellow lower back above the crimson rump. The male has a bright crimson patch on the upper breast. The female lacks red on breast, has a small blue crown patch and reduced yellow on back. Most often seen flying fast and direct high overhead on whirring wings and calling. Feeds on flowers as well as small fruits and oil palms. It roosts and often feeds upside-down.

**Where to see** Locally common resident in lowland forests, mangroves and plantations.

♀

♂

# Green Broadbill
*Calyptomena viridis* 16cm

An iridescent bright green bird with a tuft of feathers on the forehead almost obscuring the short bill. The male shows a black spot on the ear-coverts and three black bars on the wings. The female is a paler dull green, with no black markings. A relatively slow-moving bird that forages mostly in the midstorey, eating fruits, particularly figs, and some insects. The song is a quiet, unusual 'bouncing ball' trill, *boi, boi boi-boi-boi-boik*.

**Where to see** Locally common lowland resident.

# Whitehead's Broadbill  *Calyptomena whiteheadi*  26cm

Similar to Green Broadbill but larger with more black markings on the crown, throat, belly and wings. The tail is all black. The female is duller green with reduced black markings, the underparts being all green. Sometimes loud and conspicuous but often perches quietly for long periods. Usually solitary, its diet is mainly fruits and insects. The song is a raspy *ki-chrrrr*.

**Where to see** Endemic, uncommon resident in north-central montane ranges from Mount Kinabalu southwards.

# Dusky Broadbill
*Corydon sumatranus*   26cm

A remarkable, large, dark olive-brown forest bird with a huge pink bill, bare pink skin around the eyes, and a rufous throat and upper breast. In flight it shows white patches on the wings and tail. Gregarious and noisy, these broadbills forage for large insects and small vertebrates in groups of up to ten in the canopy. Easily detected by their penetrating, harsh *kweer kweer kweer kweer...* calls, which have a screaming quality.

**Where to see** Uncommon resident in lowland and hill forests.

# Black-and-red Broadbill
*Cymbirhynchus macrorhynchos*   22cm

A thrush-sized red-and-black forest bird with an upright stance and remarkable broad, bright aqua blue and pale yellow bill. Usually in pairs near flowing water, it feeds on insects and other small invertebrates, as well as small fruits. The voice is a relatively low-key rasping, chuckling *krrk krrk krrk*.

**Where to see** Common resident throughout in lowland riverine forests, mangroves and overgrown plantations.

# Black-and-yellow Broadbill
*Eurylaimus ochromalus* 14cm

Although the voice and behaviour of this bird is similar to Banded Broadbill, its appearance is very different. The bright yellow eye stands out on the black head, and the combination of the black-and-yellow back and wings with the pink breast and belly, as well as the blue bill, make this a very distinctive little forest dweller. The song is a frantic cicada-like trill, starting relatively slowly, building in pitch, volume and tempo before reaching a crescendo, then finishing suddenly.

**Where to see** Common resident throughout, in lowland to lower montane forests.

## Banded Broadbill
*Eurylaimus javanicus* 22cm

A pinkish-red bird with black upperparts with prominent yellow streaks. The wide, blue bill of this forest bird helps it catch the insects that comprise its diet. Usually in pairs, they perch quietly in the midstorey, flying out to glean prey from nearby foliage. The song is remarkable – a frantic cicada-like trill, which gradually reduces in volume, introduced by a single sudden loud *pwau*.

**Where to see** Common resident throughout in lowland to lower montane forests.

## Garnet Pitta
*Erythropitta granatina* 15cm

Very similar to Black-crowned Pitta but has a bright red crown and slightly different vocalisations. The ranges of the two species don't overlap. The call is a long, clear, slightly modulated whistle, increasing in volume and very slightly upwardly inflected, downslurred abruptly at the end.

**Where to see:** Locally common and widespread resident south of Sabah.

# Blue-headed Pitta
*Hydrornis baudii*  17cm

A shy lowland specialist, rarely found above 500m. The male has a glossy bright blue crown, blue belly and tail, and reddish-brown back. The female is very different, being reddish-brown with blue only on the rump and tail. Feeds on the ground in leaf-litter, taking caterpillars and other insects, earthworms and arthropods; sometimes seen hopping along forest trails. The alarm call is a rather loud, high-pitched, explosive *kawow*.

**Where to see** Endemic, uncommon to locally common lowland resident throughout.

# Bornean Banded Pitta
*Hydrornis schwaneri*  22cm

The pittas are among the most exciting
birds of Asia, but despite their brilliant
coloration they can be surprisingly
difficult to spot in the dense forests
that they favour. This shy electric-blue
and yellow bird can be very elusive,
especially when not calling, as it rarely
ventures into the open. The diet consists
of insects, snails and earthworms. One
call is a short, tremulous *prurr...* ; another
is a louder, strident, downslurred *pwow*.

**Where to see** Endemic, locally
uncommon resident in lowland forests.

# Hooded Pitta
*Pitta sordida* 18cm

Named for its distinctive black hood, the body is emerald-green with a bright turquoise-blue patch on the wings and a bright red patch on the belly. In flight it shows a large white patch on the wings. A lowland specialist up to 400m. Not as shy as many pittas and usually found near water. It consumes a variety of invertebrates and roosts on lianas and in low vegetation. The call is a loud, short, sharp *kwee*.

**Where to see** Uncommon resident throughout in lowland riverine forests, mangroves and plantations.

**Pittas**

# Blue-banded Pitta
*Erythropitta arquata*   16cm

This spectacular pitta has a bright orange head, dark green back and wings, and a bright red belly. The name denotes the bright turquoise-blue breast-band. It often calls from perches up to 3m above ground where it gives a low, pure whistle; it feeds on the ground on a variety of insects.

**Where to see** Rare endemic, favouring bamboo stands in hill and montane forest but is poorly known and unobtrusive as it keeps to dense undergrowth.

# Black-crowned Pitta
*Erythropitta ussheri*   15cm

The sexes are similar, with an all-black
head adorned by a thin turquoise-blue
line behind eye, a deep purplish-blue
back and wings with glossy bright blue
on the wing-coverts and a red belly.
A shy, lowland specialist that favours
thickly vegetated gullies. Its diet consists
of a variety of insects, spiders and small
snails. Usually detected by its long, clear
whistle, which increases in volume and
ends abruptly.

**Where to see** Endemic, locally common
resident confined to lowlands of N.
Borneo.

## Golden-bellied Gerygone
*Gerygone sulphurea*   10cm

The most outstanding feature of this small bird is its beautiful song, which consists of a series of glissando whistles, with a rising and falling cadence, *zwee-zii-zree-zwee-zree-zi-zeee...* A brownish bird with a golden belly, as the name suggests, which is found in all types of forest and woodland habitats. This arboreal insectivore gleans foliage in the mid- to upper storey; often joins mixed feeding flocks.

**Where to see** Common to uncommon resident throughout.

## White-bellied Erpornis   *Erpornis zantholeuca*   12cm

A small olive-green bird with a short crest, and pink legs and feet. Found in lowland to lower montane forests, it is usually solitary but sometimes joins mixed feeding flocks. Feeds on insects and less often on fruit in the midstorey.

The calls consist of various peevish, nasal buzzing notes.

**Where to see** Common resident in Sabah, uncommon in Sarawak, Brunei and Kalimantan.

# Blyth's Shrike-Babbler
*Pteruthius aeralatus* 15cm

A foliage-gleaning insectivore that is
active in the upper storey in hill and
montane forests. Usually encountered in
pairs, it sometimes joins mixed feeding
flocks. The male has a black head with
a long white supercilium from above the
eye, and chestnut patches on the glossy
black wings. The female is very different,
with an olive-grey head and back, and
an indistinct pale supercilium. The song
is a strident, loud and rather high-pitched
phrase *jip-jip...jip-ji-jip*.

**Where to see** Uncommon resident in
north-central and western mountain ranges.

♀

♂

# Dark-throated Oriole
*Oriolus xanthonotus*   20cm

A medium-sized, bright yellow songbird with a black hood and wings, and a deep pink bill. Usually solitary or in pairs, and active in the upper storey, sometimes joining mixed feeding flocks. A generalist feeder, it will take small invertebrates and figs. The song is a distinctive loud, fluty *phi-phu-phuwip*.

**Where to see** Common resident throughout in lowland forests, to 750m.

♂

♂

# Black Oriole
*Oriolus hosii*   21cm

The plumage is entirely black except the chestnut undertail-coverts, dark red eye and dull pink bill. The similar Black-and-crimson Oriole has red or chestnut on the breast and a grey bill. A poorly known bird, it forages quietly in the mid- and upper storeys of fruiting trees in montane forests, often in small parties.

**Where to see** Endemic, rare resident, known only from mountains in E. Sarawak and E. Kalimantan.

# Black-and-crimson Oriole
*Oriolus cruentus* 21cm

A glossy black oriole with a square-cut crimson patch on the lower breast and upper belly, and a glossy crimson patch on the wing. Immature birds lack red in the plumage, showing dull chestnut streaks on the lower breast and upper belly. This oriole is found in tall hill to upper montane forest where it feeds on insects, especially caterpillars, and fruit in the mid- and upper storeys, often in mixed feeding flocks.

**Where to see** Common resident in north-central montane ranges.

# Bornean Whistler
*Pachycephala hypoxantha* 16cm

One of the most commonly encountered birds in the mountains of Borneo, this medium-sized bird is olive-green with bright yellow underparts. Actively hawks and gleans insects in the mid- and upper storeys, often joining mixed feeding flocks. A variety of clear, liquid songs are given but commonly a seesawing series of eight high and low notes, *wee-sit-wee-sit-wee-sit-wee-sit*.

**Where to see** Endemic, common resident in montane forest of the north-central and western mountain ranges.

**Bristlehead**

# Bornean Bristlehead *Pityriasis gymnocephala* 25cm

One of Borneo's most iconic species, this mostly black bird with a large hooked bill is endemic to the island. Its name comes from the stiff feathers on the bare yellow-orange crown. The head is bright red with a black patch on the cheeks. In flight the wings show a white patch. The female has red spots on the flanks. It is a highly nomadic arboreal insectivore that favours lowland and hill forest where it is usually seen in small noisy flocks of six to ten in the canopy. Its voice is very varied, unusual and distinctive, with nasal contact calls and a series of high-pitched whining screams and whistles, *pit-pit-PIOU* or *pit-prree-PO*.

**Where to see** Endemic, uncommon resident throughout.

# Bar-winged Flycatcher-shrike  *Hemipus picatus*  13cm

The male is a small black-and-white forest bird with a prominent long white bar on the wings. The female has the same pattern but is blackish-brown rather than black. Sallies after flying insects and foliage-gleans in the midstorey, often returning to the same perch; usually solitary or in pairs but joins mixed feeding flocks.

**Where to see** Common resident in hill and montane forests throughout.

# Black-winged Flycatcher-shrike *Hemipus hirundinaceus* 14cm

The male's upperparts are all glossy black except for the white rump, and the underparts are white with a grey wash on the breast. The female is dark brown. Both sexes lack the white on the wings of the similar Bar-winged Flycatcher-shrike. Behaviour as in the latter species, which it usually replaces at lower altitudes, although there is overlap at middle elevations.

**Where to see:** Common resident throughout in all wooded habitats except in the mountains.

## Large Woodshrike
*Tephrodornis virgatus* 18cm

A chunky bird with a broad black mask and a large, hooked bill; in flight the white rump is noticeable. Usually in noisy pairs or small groups; generally seen high in the canopy where it feeds on large insects, hawking from exposed perches or gleaning foliage. Noisy, more often heard than seen; on Borneo the call is a harsh squeaky rattling which is quite high-pitched: *chreee-chee-chi-chi-chi.*

**Where to see** Uncommon resident in lowland and lower montane forests throughout.

# Maroon-breasted Philentoma  *Philentoma velata*  20cm

Reminiscent of a flycatcher. The male is cobalt-blue with a black face, deep maroon throat and breast, and red eyes. The female is dull cobalt-blue all over. Solitary or in pairs; usually seen in lower and midstoreys of primary forest, often in swampy areas.

**Where to see** Uncommon resident throughout.

Vangas, Helmetshrikes and Allies

# Rufous-winged Philentoma
*Philentoma pyrhoptera* 16cm

Although not a flycatcher, it does look a lot like one. The male of the most common morph is cobalt-blue with rich chestnut wings and tail, and a buffish-white belly. A rarer morph is wholly cobalt-blue, slightly paler on the lower belly and undertail-coverts, while the female is dull greyish-brown with rufous-brown wings. Usually seen singly or in pairs; a foliage-gleaning insectivore in the lower storey where it is generally inconspicuous.

**Where to see** Common resident in the lowlands throughout.

## Common Iora   *Aegithina tiphia*   13cm

This small, bright yellow bird has a dull yellowish-green head and back, and black wings with two broad white wingbars. It is usually solitary or in pairs, and gleans small invertebrates from foliage in the treetops. Occurs in degraded forests and edges, as well as mangroves, plantations and gardens.

**Where to see** Common resident throughout.

# Green Iora
*Aegithina viridissima*   12cm

Similar to Common Iora but darker green
with a greenish-yellow belly, black wings
with two broad white bars, and a broad,
broken, bright yellow eye-ring. Gleans
foliage for small invertebrates at all
levels, often joining mixed feeding flocks.
Found in disturbed areas of lowland
forest, forest edges, mangroves and
overgrown plantations.

**Where to see** Common resident
throughout.

# White-breasted Woodswallow *Artamus leucorynchus* 18cm

A gregarious bird of open areas near mangroves, plantations and paddyfields. It catches insects in flight with its feet and eats them on the wing. Often seen in small groups huddling together on powerlines and bare branches. They characteristically wag their fanned tail from side to side when perched.

**Where to see** Common resident throughout.

## Sunda Cuckooshrike
*Coracina larvata* 22cm

A slaty-grey bird with a black mask and black wings with grey fringing. The female has a reduced dark mask. Found in hill to upper montane forest where it feeds in the upper storey on fruit, large invertebrates and small vertebrates such as geckos. Noisy and conspicuous; usually solitary or in pairs.

**Where to see** Uncommon resident in the north-central mountain ranges.

# Lesser Cuckooshrike
*Coracina fimbriata*   17cm

Similar to Sunda Cuckooshrike but smaller with darker blackish-grey wings and tail, and no black mask. The female is quite different with a barred face and underparts. This cuckooshrike feeds on small invertebrates and fruit in the upper storey, sometimes joining mixed feeding flocks.

**Where to see** Common resident throughout in primary and secondary lowland, peat swamp and hill forests.

# Pied Triller
*Lalage nigra*   16cm

A black-and-white bird of open country, most often seen in lowland coastal woodlands, mangroves, cultivated areas and gardens. Generally forages for insects in the foliage of the upper storey but will come to the ground; usually solitary or in pairs. The male is black with a white supercilium, pale belly and large white patch on the wings. The female is similar to the male but tinged brown.

**Where to see** Common resident throughout.

## Scarlet Minivet
*Pericrocotus speciosus* 18cm

Similar to Fiery Minivet but larger, lacking the orange tinge, with more red on the tail and showing a second, smaller wing-patch. The male is red and black, while the female is yellow with a grey head and back. An arboreal insectivore, usually in active flocks that flit from treetop to treetop in the upper storey.

**Where to see:** Locally common resident throughout, possibly commoner in N. Borneo (Sabah).

# Fiery Minivet
*Pericrocotus igneus*  15cm

The male is a striking red-and-black bird, while the female is yellow with a dark grey crown and back. An arboreal insectivore; usually seen in active flocks flitting from treetop to treetop in primary and secondary lowland forest, woodlands and plantations; a lowland specialist.

**Where to see** Uncommon lowland resident throughout.

# Grey-chinned Minivet
*Pericrocotus solaris*  17cm

Similar to the other minivets but the male Grey-chinned Minivet has a dark grey throat and the female lacks yellow on the face, and the wing pattern of both is different. An arboreal insectivore, it often joins mixed feeding flocks. Found at higher elevations than other minivets.

**Where to see** Common montane resident in north-central ranges from Mount Kinabalu southwards.

# White-throated Fantail   *Rhipidura albicollis*   19cm

A bold, active, black-and-white montane bird with a distinctive fan-like tail. This species is a very active insectivore that catches insects on the wing generally in the lower to midstorey. It often spreads its tail, fanning it out in a jaunty manner to scare up insect prey; joins mixed feeding flocks. The song is a pretty series of four to six thin, high-pitched notes, a descending then rising *dee-dee-dee-dit-dooo*.

**Where to see** Common resident in the mountains throughout.

# Malaysian Pied Fantail   *Rhipidura javanica*   18cm

Similar to White-throated Fantail but has white underparts and is found in lowland habitats. A very active insectivore, constantly fanning its tail when flycatching and sallying while feeding in the lower to midstorey. The voice is a high-pitched, squeaky *wi-tit-weet-weet-weet-wit*.

**Where to see** Common lowland resident throughout, in open forests, secondary growth, plantations, cultivated land and gardens.

# Spotted Fantail  *Rhipidura perlata*  17cm

A dainty black-and-white fantail with distinctive spotting on the breast. Similar to other fantails but with a string of small white spots on the wing-coverts, throat and breast. Unlike the other two species on Borneo, it favours old-growth forest. An active insectivore; usually seen singly or in pairs in the mid- to upper storey; tends to perch with an upright stance.

**Where to see** Common resident throughout in lowland forests.

# Ashy Drongo
*Dicrurus leucophaeus* 27cm

A pale grey drongo with a prominent white spot in front of the eye, a long deeply forked tail and a red eye. Hawks for insects from a prominent high perch, usually at the edge of an open area; one of the most visible birds at higher altitudes.

**Where to see** Common resident in hill and mountain ranges in north; less common in south.

# Bronzed Drongo
*Dicrurus aeneus* 22cm

An all-black, glossy drongo with a forked tail that lacks upturned tips, and a dark red eye. Sallies out in pursuit of insects from shaded perches in the mid- to upper storey but favours open areas within the forest.

**Where to see** Common resident throughout in lowland, riparian and swamp forests.

# Hair-crested Drongo  *Dicrurus hottentottus*  31cm

Similar to Bronzed Drongo but with a crest of hair-like plumes on head (often difficult to see), a broad, almost unforked tail prominently curved up and out at the tips, and a strong dark greenish gloss on the wings and tail. Noisy and conspicuous as it hawks insects from prominent perches in clearings. Vocalises with highly varied loud chirps, peeps, whistles and rasping notes.

**Where to see** Common resident throughout in hill to lower montane forest.

# Greater Racket-tailed Drongo
*Dicrurus paradiseus*   30cm (up to 60cm with rackets)

A large, all-black drongo with glossy dark blue upperparts, a prominent short crest on forehead, and a shallow-forked tail with highly elongated outer feathers (up to 30 cm), with bare shafts and twisted terminal rackets (sometimes missing). Often in mixed feeding flocks, it is noisy and conspicuous, and more often encountered within forest than other drongos. The voice consists of highly varied loud rasping, squeaking and churring notes, with some whistles and mimicry.

**Where to see** Common lowland forest resident throughout.

# Black-naped Monarch
*Hypothymis azurea*   16cm

A small, active, noisy insectivore with a short crest that gives it a square-headed appearance. The male is azure-blue with a black spot on the top of the head, a black stripe across the upper breast and a blue bill. The female is duller with a pale blue head. Usually seen singly or in pairs foraging in the lower to midstorey. The call is a wheezy *shweet shweet...*

**Where to see** Common resident throughout in lowland to lower montane forest.

# Blyth's Paradise-Flycatcher

*Terpsiphone paradisi* 20cm (excluding male's long tail feathers)

This striking flycatcher occurs in two morphs, white and rufous. On Borneo the latter, which has a rufous back, is very rare; the white form dominates. The male of the white form has a glossy black head (with a short crest) and throat, and the rest of the plumage is white. The elongated tail can be up to 30cm long, and the blue eye-wattle and bill are notable. The female is rufous with a short tail. This arboreal insectivore hawks insects from an inconspicuous perch.

**Where to see** Common resident throughout in lowland to lower montane forest.

# Crested Shrikejay *Platylophus galericulatus* 27cm

A remarkable tawny-brown bird with a large white patch on the sides of the neck and a long erect crest. Usually encountered in small parties, it is noisy but often quite elusive in the dense vegetation of the midstorey of dense forest where it gleans small to large invertebrates. The song is a distinctive loud machine-gun rattle, *tit-tut-tut-tut-tut-tut...*

**Where to see** Locally common resident throughout in lowland to lower montane forests.

# Malaysian Rail-babbler

*Eupetes macrocerus*   28–30cm

This distinctive rail-like ground-dweller is very shy and easily overlooked. The first clue to its presence is usually the long, clear, whistled song delivered on a flat pitch and increasing in volume, very like that of Black-crowned and Blue-banded Pittas but monotone, higher and shorter. This long-necked rufous bird can be seen pursuing invertebrate prey in dense vegetation on the forest floor; it rarely flies. The head pattern is striking, with a bold bright white supercilium from in front of the eye to the sides of neck, a conspicuous broad black stripe through and below the eye from the lores to the sides of neck, and a thin patch of bright blue skin running along the neck-sides below the eye-stripe (sometimes inflated when calling).

**Where to see** Rare resident in primary lowland dipterocarp and hill forests, probably confined to N. and C. Borneo.

**Crows, Jays and Magpies**

# Bornean Green Magpie   *Cissa jefferyi*   32cm

A very striking bright green bird with maroon-red wings, a black mask, and bright pinkish-red bill and legs. It often joins mixed feeding flocks in the lower to midstorey of tall forest, usually in small parties. They are noisy but often quite elusive. The voice consists of many varied notes and phrases, including a sweet, piping *wit-wit-wit-tswit-tswit-wu-wooo*.

**Where to see** Endemic, common resident in north-central mountain ranges.

# Bornean Black Magpie
*Platysmurus aterrimus* 37cm

An all-black bird with a short erect crest on the forehead, a long broad tail and red eyes. These noisy and active foliage-gleaning insectivores of the midstorey are often in pairs but sometimes in quite large parties. The song is a long, monotone, mellow, repeated whistle with the quality of a bamboo flute, *whooo... whooo...* A lowland specialist in primary rainforests and swamp forests.

**Where to see** Local and uncommon lowland resident throughout.

# Bornean Treepie
*Dendrocitta cinerascens* 40cm

A long-tailed fawn-brown bird with a grey crown, back and tail, and black wings. It is a generalist feeder in hill to upper montane forests, where it forages for insects and fruits. Noisy and conspicuous, it is usually seen singly or in pairs. The call is a repeated harsh, grating *krek-krek...*

**Where to see** Endemic, locally common resident in north-central and south-western mountain ranges.

# Slender-billed Crow  *Corvus enca*  45cm

A typical crow, similar in appearance to many other crows but the only one that is widespread and regularly seen on the island. Compared with the others, Slender-billed Crow has a squarer tail, shorter, more rounded wings, and faster wingbeats with the wings held below the horizontal in flight. When perched it has a more slender build and thinner bill with a less arched culmen. Relatively shy, it avoids human habitation.

**Where to see** Common lowland resident throughout.

# Grey-headed Canary-Flycatcher  *Culicicapa ceylonensis*  13cm

A small olive-green flycatcher with a grey head and yellow belly. The head has a slight crest, giving it a square-headed look. Not shy; usually solitary, it actively forages for insects and other small invertebrates at all levels, often flicking its tail when perched. The song is a squeaky, loud, constantly repeated *sil-ly-bil-ly-me*.

**Where to see** Common resident throughout.

# Barn Swallow  *Hirundo rustica*  16cm

This familiar bird is found in open country including around cities and towns, sometimes in huge numbers where they can be seen foraging for small insects in flight over open ground or fresh water. The glossy blackish-blue upperparts contrast with the creamy-white underparts and brick-red forehead and throat, bordered below with a blackish-blue breast-band. The forked tail with elongated outer streamers is distinctive.

**Where to see** Non-breeding visitor and passage migrant from July to May.

# Pacific Swallow  *Hirundo tahitica*  12cm

Though similar to Barn Swallow, this species lacks the breast-band and long tail-streamers. The underparts are more greyish-white and the tail has only a shallow fork. Pacific Swallow can be seen in any open area including in primary and secondary forest, freshwater wetlands, estuaries, beaches, paddyfields and towns. It is an aerial insectivore, foraging for small insects over open ground or fresh water.

**Where to see** Common resident throughout.

# Straw-headed Bulbul
*Pycnonotus zeylanicus* 28cm

This large bulbul is famous for its rich
and melodious, powerful, far-carrying
song which, with luck, can be heard in
riverine vegetation in lowland forests.
It forages in low bushes near water,
feeding on small fruits and berries, and
occasionally small invertebrates. The
golden-yellow head with a narrow
black eye-stripe and broad malar stripe
immediately identifies this otherwise
greyish-olive bulbul.

**Where to see** Uncommon resident
throughout, formerly common but now
rare owing to cagebird trade and, to a
lesser extent, habitat loss.

# Black-headed Bulbul
*Pycnonotus atriceps* 17cm

A smaller, olive-green bulbul with a
black hood, pale blue eye, bright yellow
lower back, and tail tipped yellow with a
broad black subterminal band. Like many
frugivorous birds, this bulbul is nomadic
as it searches for fruiting trees in lowland
to lower montane forests. The song is a
rapidly repeated *doo-da-dit-dit-doo*.

**Where to see** Large fluctuations in
abundance but generally common
resident throughout.

# Bornean Bulbul
*Pycnonotus montis*  18cm

A mountain bird that is easily identifiable by its black head and crest contrasting with the olive-green back and yellow underparts. An arboreal frugivore, it also takes flying insects. Usually solitary or in pairs but small groups congregate at fruiting trees. The song is a loud, sweet, emphatic *wee-wit-it-weet*.

**Where to see** Endemic, uncommon in north-central mountain ranges from Mount Kinabalu to Ulu Barito and Mount Menyapa.

# Scaly-breasted Bulbul
*Pycnonotus squamatus*  15cm

One of the prettiest and most distinctive bulbuls with a scaly pattern on the underparts and a black head contrasting with a white throat and bronze-green back. Usually in pairs or small groups in lowland and hill forests. An arboreal foliage-gleaning frugivore that occasionally takes small insects, it is known to hover at the edge of foliage. These bulbuls gather at fruiting trees, foraging in the forest canopy and edges.

**Where to see** Uncommon resident in hills and mountain ranges throughout.

## Grey-bellied Bulbul
*Pycnonotus cyaniventris* 17cm

A dark smoky-grey bulbul with a short pale supercilium, olive-green back and wings, and bright yellow undertail-coverts. A midstorey foliage-gleaning frugivore and insectivore that inhabits lowland and lower montane forest. It is an inconspicuous bird usually seen in mixed feeding flocks.

**Where to see** Uncommon resident throughout.

## Olive-winged Bulbul
*Pycnonotus plumosus* 20cm

A rather nondescript brownish bird with olive-green fringes on the wings and tail, faint whitish streaks on the ear-coverts and a red iris. It feeds on berries, small fruits and small invertebrates in lowland forest, scrub and mangroves.

**Where to see** Common resident throughout.

# Pale-faced Bulbul
*Pycnonotus leucops*   18cm

This bulbul is distinguished by its pale face and lores, giving it a wide-eyed appearance. The rest of the plumage is a dull olive-brown with bright yellow undertail-coverts. It inhabits montane forest and heath where it is usually seen in pairs or small groups. Mainly feeds on small fruits. The song is a bubbly, emphatic *kw't-kw't tu-weeti-weeti*.

**Where to see** Endemic, locally common resident in north-central mountain ranges from Mount Kinabalu to Mount Murud and Mount Menyapa.

# Yellow-vented Bulbul *Pycnonotus goiavier* 20cm

An olive-brown bulbul with a short, narrow, dark olive-brown crest, and broad white supercilium and face with brownish ear-coverts. The underparts are whitish with brownish smudges on the breast and yellow undertail-coverts. Usually seen in secondary forests, mangroves, plantations, cultivated areas and gardens where it feeds on small fruits and berries, and small invertebrates. The song is a cheery, bubbly *pirit pirit pirit pirit prrt...*

**Where to see** Abundant resident throughout.

# Red-eyed Bulbul
*Pycnonotus brunneus* 19cm

Another nondescript bulbul. It has olive-brown upperparts with paler olive fringing on the wings and buffish undertail-coverts; the iris has a red outer ring, and the legs and feet are pinkish-brown. The similar Cream-vented Bulbul is slightly smaller with paler underparts and a variety of iris colours from white to orange or red, which can make positive identification difficult. Both species are found in secondary to lower montane forests, mangroves and plantations.

**Where to see** Common resident throughout.

# Spectacled Bulbul  *Pycnonotus erythrophthalmos*  18cm

The most distinctive feature of this bulbul is, as the name suggests, its red iris with a narrow yellow eye-ring, as well as its yellowish gape. Otherwise, it is an olive-brown bird with a rufous tinge on the rump and uppertail-coverts, and pinkish-brown legs and feet. It can be separated from Red-eyed and Cream-vented Bulbuls by the eye-ring and yellow underwing-coverts. The song is a sweet, complex series of relatively high-pitched notes *pi-pi-pi-dee-dee-dee...*

**Where to see** Common resident in primary and secondary to lower montane forests, mangroves and overgrown plantations throughout.

## Ochraceous Bulbul
*Alophoixus ochraceous*  21cm

A chunky rufous-brown bulbul with a shaggy crest, a white throat and rich rufous-brown undertail-coverts. Grey-cheeked Bulbul is similar but has a less prominent crest, yellowish underparts and buffy undertail-coverts. Ochraceous Bulbuls are conspicuous and noisy in lower and upper montane forest. The song is a rapid series of raspy *kchit-kchit-phew-phew-phew-phew-phew* notes.

**Where to see** Common resident in mountain ranges throughout.

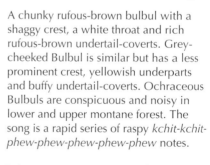

# Hairy-backed Bulbul
*Tricholestes criniger* 17cm

A somewhat flycatcher-like bulbul with a pale yellow face, olive-green upperparts and yellow underparts. The namesake long hair-like plumes on the back are rarely visible in the field. It can be detected by its single clean whistle call, *phweee;* the song is a series of rather quiet, scratchy, jumbled notes. Forages for small fruits in the lower to midstorey of primary and secondary lowland forests, and overgrown plantations.

**Where to see** Common resident throughout.

# Charlotte's Bulbul
*Iole charlottae* 20cm

A drab bulbul that is best distinguished by its buffish undertail-coverts, which contrast with the greyish underparts and warm brown upperparts. They forage at all levels for small fruits and invertebrates in primary and secondary lowland forests. The call is a rapidly repeated nasal *chiwit.*

**Where to see** Endemic, uncommon resident throughout.

# Cinereous Bulbul
*Hemixos cinereus* 20cm

A distinctive bulbul with a dark face and moustachial area contrasting with the white throat, and a shaggy crest. The upperparts are olive-brown with greenish fringing on the wings and tail, and a bright yellow vent. This arboreal insectivore and frugivore often gathers at fruiting trees in hill forests where it is noisy and conspicuous.

**Where to see:** Common resident in mountain ranges throughout.

# Hook-billed Bulbul
*Setornis criniger* 20cm

This unusual bulbul has a unique shrike-like hook-tipped bill. The upperparts are dark brown, and the bird has a short whitish supercilium and black eye-stripe and malar stripe. The underparts are white with a grey wash on the flanks and a buffish wash on the lower belly. Occurs only in lowland coastal peat swamp and heath forests. It is usually seen in pairs or small groups foraging for small fruits and invertebrates in the lower to midstorey.

**Where to see** Scarce to locally common resident with a patchy distribution, rarer in the north and absent from the east.

# Bold-striped Tit-Babbler
*Mixornis bornensis* 12cm

A chestnut-brown babbler with boldly streaked underparts, this very noisy and active arboreal insectivore is usually found in small parties foraging for invertebrates in dense vegetation in disturbed swamp and riparian forest, mangroves, scrub, cultivated areas, gardens and plantations. The pale yellow iris is surrounded by blue orbital skin. The paced, mellow *chonk-chonk-chonk-chonk...* song of the male is instantly recognisable and is usually accompanied by a constant *chk-chk-chk-chk-chk...* uttered by the female.

**Where to see** Abundant resident throughout.

# Fluffy-backed Tit-Babbler
*Macronus ptilosus* 17cm

This charismatic chestnut-brown babbler has a rich rufous crown and black throat, as well as bright blue skin around the eye. When singing, the pale blue skin on the neck-sides often inflates like a little balloon. The eponymous long plumes on the rump and flanks are usually not visible. These babblers are typically found in pairs or small parties foraging in dense vegetation in the lower storey of disturbed areas of lowland forests. The song is a pleasant, low-pitched, mellow *poop...poop-poop-poop,* usually with a strange, very throaty churring, probably uttered by the female.

**Where to see** Locally common resident throughout.

# Chestnut-winged Babbler *Stachyris erythroptera* 13cm

A charming and noisy dark grey babbler with a chestnut-brown back and wings. When singing the inflated blue skin shows on the neck-sides. This active and noisy foliage-gleaning insectivore is usually seen in small parties in thick vegetation in the midstorey, often in mixed feeding flocks.

The mellow, slow-paced *hu-hu-hu-hu-hu* song is often accompanied by churring notes, probably uttered by the female.

**Where to see** Common resident throughout in lowland forests, mangroves and plantations.

# Rufous-fronted Babbler
*Stachyridopsis rufifrons* 12cm

An inhabitant of dense undergrowth in lowland and hill forests, this little olive-brown babbler with a chestnut crown is usually seen in pairs or small groups foraging in dense foliage for small invertebrates. It often joins mixed feeding flocks and can be heard uttering a simple mellow song, *pu pu-pu-pu-pu-pu-pu-pu...* with a pause after the first note.

**Where to see** Uncommon resident in Sabah, Sarawak, and W., C. and E. Kalimantan.

# Bare-headed Laughingthrush  *Garrulax calvus*  26cm

This very unusual blackish-brown bird is instantly recognisable by the bare yellowish skin on its forehead and crown, giving it a bald appearance. The bill is bright orange-red, framed by a blue submoustachial skin patch. It is often an elusive bird, usually found in pairs or small flocks foraging for small invertebrates, creeping about in columns of dense vegetation in the midstorey. The calls are a very low, soft, hollow series of 3–20 short notes, *hoo hoo hoo hoo...*

**Where to see** Endemic, locally uncommon resident in montane forest in the north-central mountain ranges.

Tree-Babblers, Scimitar-Babblers and Allies

## Chestnut-backed Scimitar-Babbler *Pomatorhinus montanus* 21cm

A highly distinctive but shy babbler whose most conspicuous feature is its longish decurved yellowish bill. The head is black with a prominent long white supercilium, bright chestnut upperparts and white belly. The low-pitched four-note song, *grrt-put-put-pweet* of which the first note is a short, quiet growl, is usually the first clue to its presence. It probes bark and foliage in the mid- to upper storey for invertebrate prey in lowland to lower montane forests.

**Where to see** Uncommon resident throughout.

## Grey-throated Babbler *Stachyris nigriceps* 14cm

This small brown babbler with a black crown and short pale grey supercilium behind the eye has a blackish throat with prominent white submoustachial patches. It is a familiar sight and sound in hill and montane forests where it is usually found in small, noisy and active groups of five to eight individuals in the lower storey. Its song is a rapid, high-pitched monotone, *tsi-tee-ti-ti-ti-ti-ti-ti-ti-ti*.

**Where to see** A common resident in the north-central, western and southern mountain ranges.

# Black-throated Babbler *Stachyris nigricollis* 16cm

An active babbler that is usually seen at forest edges in the lowlands. It feeds by gleaning insects from the foliage in the lower to midstorey, usually in pairs or small parties. The rufous-brown body with black throat and prominent broad white malar spots, along with its mellow, fluty *pu pu-pu-pu-pu-pu-pu-pu-pu-pu-pu-pupu-pu...* song, make it relatively easy to identify.

**Where to see** Locally common lowland resident throughout.

# Chestnut-rumped Babbler *Stachyris maculata* 18cm

A sometimes-elusive babbler of the lowlands with a cheery, low-pitched hooting *woop-oop-woop...* song. An active and noisy foliage- and bark-gleaning insectivore, it is usually found in small parties in dense vegetation in the lower to midstorey, and it often joins mixed feeding flocks. The body is olive-brown with a bright chestnut lower back, while the white breast to belly is heavily streaked. When singing, a patch of inflated blue skin can be seen on the neck-sides.

**Where to see** Common resident in lowland forest throughout.

# Black-throated Wren-Babbler *Turdinus atrigularis* 18cm

With great care, this ground-dwelling insectivore can be seen gleaning leaf-litter and flicking leaves on the forest floor in densely vegetated gullies. It can be identified by its black throat to upper breast, and buffish underparts with broad black scaling. The back is brown with broad black fringing, and there is a patch of blue skin behind the eye. The bird is usually heard before it is seen, singing a loud, crowing monotone, *dit-dee-dooo-doooo*.

**Where to see** Endemic. An uncommon lowland resident.

# Moustached Babbler *Malacopteron magnirostre* 17cm

An active olive-brown tree babbler with a dark grey crown, a narrow dark grey moustachial stripe and a pale grey wash on the breast. It often joins mixed feeding flocks with other *Malacopteron* babblers including the similar Sooty-capped Babbler, which has a better-defined, browner crown, a darker tail and no moustachial stripe. The song is a loud, melancholy three- to six-note *di-doo-doo-doo-dooo*.

**Where to see** Common resident throughout in lowland forests.

# Sooty-capped Babbler *Malacopteron affine* 16cm

Another plain tree babbler with an olive-brown back and greyish-brown cap, chestnut tail and pale grey wash around the eye. It is found in lowland forests where it favours secondary growth and areas of natural disturbance. The pleasant song is a very distinctive rambling series of rather slow, plaintive, clear whistles, *phu-phi-phu-phoo-phi-phoo-phu-phi...*, seemingly randomly rising and falling in pitch.

**Where to see** A common lowland resident throughout.

# Scaly-crowned Babbler *Malacopteron cinereum* 15cm

An olive-brown tree babbler with a black-speckled rufous nape and paler underparts with a whitish throat. The very similar Rufous-crowned Babbler is larger, with grey streaking on the throat and breast. All these rather similar tree babblers are most easily differentiated by their songs. This species has a melancholy clear whistled *dui-dii-doo...* on a rising scale, the first note slightly dipping. It is an arboreal insectivore, active, acrobatic and conspicuous in the mid- to upper storey; usually encountered in small parties.

**Where to see** Common resident throughout in lowland forests and occasionally in old plantations.

# Rufous-crowned Babbler *Malacopteron magnum* 19cm

Another olive-brown tree babbler differing from the previous species by its plain rufous cap, black nape and white underparts with broad grey streaks on a greyish-washed breast. It also lacks the black tips to the crown feathers and has bluish-grey legs. The song is a slightly complaining whistled series of notes on a descending scale. It also has a wandering song similar to Sooty-capped Babbler but slightly slower-paced with longer notes.

**Where to see** Common in lowland forests and old plantations.

# Black-capped Babbler *Pellorneum capistratum* 16cm

This small babbler can be found on the forest floor searching the leaf-litter for its insect prey. Although it is somewhat secretive and inconspicuous, it can be detected by its clear sweet whistled *pii-yuu*, which rises slightly then falls in pitch. Along with its horizontal posture, the bird's chestnut-brown back, black head with long whitish supercilium and white throat make the bird quite distinctive.

**Where to see** Common resident throughout in lowland forests.

# White-chested Babbler *Pellorneum rostratum* 16cm

One of the main features of this rather nondescript brown babbler is its behaviour. It is a riverbank specialist, usually found near water where it forages on the ground, rocks and roots at the water's edge. Also look for the white underparts with pale grey sides to the breast and flanks. The similar Ferruginous Babbler is brighter rufous, has a longer tail, and has different behaviour and habitat.

**Where to see** Locally common resident throughout in lowland riverine forests.

# Ferruginous Babbler *Pellorneum bicolor* 17cm

The rufous plumage of this tree babbler is distinctive. It is often quite inconspicuous in the mid- and lower storey of the forest, where it forages for insects, especially ants; it often joins mixed feeding flocks. The song is a simple, short, high-pitched, sharply upslurred *hweet*.

**Where to see** A locally common resident throughout in lowland forests.

# Temminck's Babbler  *Pellorneum pyrrogenys*  15cm

This small babbler with warm brown plumage, a dark grey head and white throat can be found in hill and lower montane forests where it forages on or near the ground in thick vegetation. The song is a loud jolly whistled *pi-choo,* uttered irregularly, the first note upslurred and the longer second note downslurred.

**Where to see** An uncommon resident throughout.

## Short-tailed Babbler *Pellorneum malaccense* 14cm

**Ground Babblers and Allies**

This small terrestrial insectivore with warm brown plumage and a distinctive head pattern can be seen hopping on the ground searching the leaf-litter for its prey. The face is grey with a narrow blackish moustachial stripe and white throat, and it has a pale tawny-rufous wash on underparts. It could be mistaken for the larger Abbott's and Horsfield's Babblers, which both lack the blackish moustachial stripe and have longer tails. The mournful song consists of a series of quite long, descending whistles, *phwee-phwee-phwee-phwee...*

**Where to see** Locally common in lowland forests throughout.

## Striped Wren-Babbler
*Kenopia striata* 14cm

This small ground-dwelling babbler has distinctive chestnut-brown plumage with broad white streaks and a white face with yellow bristly lores. The bird's underparts are white with grey mottling on the breast-sides and chestnut-streaked flanks. The song, which is the best clue to its presence, is an even-pitched clear and pleasant, *doo-tee-doooo*. Usually encountered singly or in pairs hopping on the ground, gleaning small insects from leaf-litter in thick vegetation.

**Where to see** Locally uncommon to rare lowland resident throughout.

# Mountain Wren-Babbler  *Gypsophila crassa*  14cm

This terrestrial insectivore favours densely vegetated rocky gullies and is usually encountered in noisy, often inquisitive family groups of four to six. The plumage is mostly dark greyish-brown with buffish streaks, and the bird has a whitish-grey supercilium. The typical song is a jolly *pee-tee-pee-ti-pee-ti*. The similar Eyebrowed Wren-babbler is smaller, paler and buffier on the throat and breast, with diagnostic buffish-white spots on the wing-coverts.

**Where to see** Endemic, locally uncommon to common resident in primary lower and upper montane forest in the north-central and western mountains.

# Bornean Wren-Babbler *Ptilocichla leucogrammica* 16cm

This secretive and shy ground-dwelling insectivore can be seen gleaning leaf-litter in densely vegetated gullies. It moves like a small rail and is best detected by its distinctive song, a clear, sweet mournful *doo-dee* or *doo-dee-doo*. Although not brightly coloured, it is an attractive bird with chestnut-brown upperparts and contrasting scaly whitish and grey sides to the head.

**Where to see** Endemic, a locally uncommon lowland resident throughout.

# Brown Fulvetta *Alcippe brunneicauda* 15cm

This small, very nondescript, arboreal babbler, found in lowland to lower montane forest, is brown with a dull greyish-brown crown and pale grey face and underparts. It gleans foliage for small insect prey and fruit in the midstorey, often joining mixed feeding flocks. The species is easily overlooked, and usually detected by its distinctive song, a loud, sweet, up-and-down *di-ti-di-ti-du-dit*.

**Where to see** Common resident throughout.

# Chestnut-hooded Laughingthrush *Ianthocincla treacheri* 23cm

One of the most conspicuous of the mountain birds, this noisy laughingthrush can be seen foraging from the ground to the upper storey for small invertebrates and fruits; usually in small groups, often joining mixed feeding flocks. The bird is grey with a chestnut hood, white wing patch, bright yellow crescent below the eye and orange bill.

**Where to see** Endemic, common resident in north-central ranges in forest, forest edges and cultivation.

# Sunda Laughingthrush *Garrulax palliatus* 25cm

These noisy and charismatic birds can often be heard in the hill forests of Borneo. They are instantly recognisable by the slaty-grey head and breast, which contrasts with the otherwise chestnut-brown plumage. The black eye is surrounded by a broad aquamarine eye-ring. Small loose groups of birds can be heard singing a low, mellow, ringing *koo-koo-koo-koo...* reaching a rapid high-pitched chattering crescendo, *wit-wit-wit-witty-witty...* They forage for small invertebrates and berries at all levels, often in mixed feeding flocks.

**Where to see** A common resident in the north-central montane ranges.

# Chestnut-crested Yuhina  *Yuhina everetti*  14cm

This nomadic frugivore with warbler-like movements can be seen in hill and mountain forests where it forms fast-moving flocks. They actively feed in the forest canopy, flitting rapidly from tree to tree uttering an unobtrusive, high-pitched *tsee tsee tsee*. This species' most conspicuous feature its rich chestnut peaked crest, which contrasts with the brown back and greyish-white underparts.

**Where to see** Endemic, uncommon in north-central and western mountain ranges.

# Black-capped White-eye
*Zosterops atricapilla* 10cm

Usually seen in flocks, this small olive-green bird often joins mixed feeding parties in the lower and upper montane forests. It has a black forehead and lores, and a broad bright white eye-ring. The similar Oriental White-eye lacks black on the face, has darker upperparts and paler grey underparts with a narrower, longer yellow ventral stripe.

**Where to see** Common resident in north-central mountain ranges.

# Javan White-eye
*Zosterops flavus* 10cm

An all-yellowish-green white-eye with green fringes on the tail and no black on the face. The similar Lemon-bellied White-eye has black lores. Javan White-eye is found in mangroves and coastal scrub at sea level where it gleans small invertebrates from treetop foliage.

**Where to see** A common resident in south-east Borneo.

# Mountain Black-eye
*Chlorocharis emiliae* 12cm

This little dark olive-green bird is distinguished by the small black area on the lores and around the eye, which is bordered with bright yellowish-green, and its long, pointed pink bill. It feeds on small invertebrates and nectar at all levels of the forest, often in small groups. It sings a sweet, melodious, thrush-like *wit-weet-weet-weet-weetee-weetee-tee.*

**Where to see** Endemic, a common resident in the north-central and western mountains.

# Mountain Leaf Warbler *Phylloscopus trivirgatus* 11cm

Commonly seen in mixed feeding flocks, this small warbler gleans insects in the foliage of the mid- to upper storey of mountain forests. A small greenish-grey bird with an indistinct paler crown-stripe and broad yellowish supercilium, it is inconspicuous but its melodious whistle, *tsee-wi-chi-wi-chi-wi-chit*, is often heard.

**Where to see** Common montane resident in north-central, western and southern ranges.

# Yellow-breasted Warbler *Seicercus montis* 9–10cm

A distinctive olive-green warbler with a chestnut crown and face patterned with black lateral crown-stripes, two yellow wing-bars and bright yellow throat and underparts. The conspicuous white eye-ring and pinkish bill and legs are also notable. They commonly join mixed feeding flocks, gleaning small invertebrates from the foliage in the mid- to upper storey.

**Where to see** Common resident in montane forests in the north-central and western ranges.

# Yellow-bellied Warbler *Abroscopus superciliaris* 9cm

A foliage-gleaning insectivore, usually seen singly, foraging in thick undergrowth and bamboo stands in lowland to lower montane forests in the lower to midstorey. The plumage is olive-green with a prominent greyish-white supercilium and lemon-yellow underparts. The song is a pleasant piercing tinkling series of six to eight descending notes, *di-dee-dee-dee-dee-dee-dee-dit*.

**Where to see** Common resident throughout, rare in Brunei.

# Sunda Bush Warbler *Horornis vulcanius* 13cm

The very distinctive song of this otherwise inconspicuous chestnut-brown warbler is a pleasant, slurred phrase of four short notes followed by a long note, that rises and falls noticeably: *wit-w't-a-wit weeee-ik*. It inhabits dense undergrowth in lower and upper montane forests, skulking in thickets, where it gleans small invertebrates, especially near road cuttings.

**Where to see** Common resident in north-central mountain ranges.

# Bornean Stubtail *Urosphena whiteheadi* 10cm

This remarkable little bird is an upright, brownish ground-dwelling warbler with a long buffish-brown supercilium and cheeks, and pale pink legs and feet. It can be seen hopping and creeping unobtrusively, though often confidingly, on the ground or in low undergrowth in the lower and upper montane forests of the north-central mountain ranges. The song is a short, very high-pitched single note, inaudible to many people.

**Where to see** Endemic, uncommon resident in N. Borneo.

# Mountain Tailorbird  *Phyllergates cucullatus*  11cm

This colourful tailorbird's sweet, musical *dee dee-dee-dee-di'di'di* whistles are often heard in the montane forests of Borneo. It is a rather skulking olive-green bird with a bright orange-rufous crown, grey head and bright yellow belly. It gleans tiny invertebrates from foliage in thickets, bamboo stands and dense vegetation.

**Where to see** A common resident in the north-central and western mountain ranges.

# Dark-necked Tailorbird
*Orthotomus atrogularis*  12cm

The male of this perky bird, with a distinctive cocked tail, is olive-green with a rufous crown and blackish-grey throat and upper breast. The female is paler rufous on the head and lacks the dark breast. The jaunty reeling *prreee* song can be heard in lowland forest where it favours disturbed areas and forest edges.

**Where to see** A common resident throughout.

# Rufous-tailed Tailorbird *Orthotomus sericeus* 13cm

The pleasing, rollicking *do-dee do-dee do-dee* continuously repeated song of this tailorbird is often heard at the edges of lowland to lower montane forests, including in secondary forest, plantations, overgrown gardens and mangroves. With its cocked tail and distinctive bright rufous-chestnut head and tail contrasting with otherwise grey plumage, this bird is easily identified.

**Where to see** A common resident throughout.

# Ashy Tailorbird *Orthotomus ruficeps* 12cm

A dark ashy-grey tailorbird with a deep rufous hood, and pink bill and legs. The female is paler on the underparts with a more washed-out hood. An inhabitant of all forest types in the lowlands, including plantations and gardens, where it gives a variety of loud, strident songs and calls including a relatively slow, upwardly inflected *choowit...choowit...*

**Where to see** A common resident throughout.

# Velvet-fronted Nuthatch *Sitta frontalis* 13cm

This striking, but often inconspicuous, violet-blue nuthatch, with its bright red bill and eye-ring, is usually seen creeping on the trunks and large branches of trees where it searches for small invertebrates. It is often seen in mixed feeding flocks in lowland hill forests. The subspecies found on Borneo has bright pinkish-red legs and feet.

**Where to see** Common resident throughout.

# Yellow-bellied Prinia *Prinia flaviventris* 14cm

The charming but repetitive *wit-wit weety-weety-weet* burbling warble of Borneo's only prinia is often heard emanating from freshwater wetlands, grasslands, paddyfields and roadside scrub in the lowlands and hill forests. It feeds on small invertebrates in dense vegetation close to the ground, sometimes singing from low bushes and tops of tall grass stems. Despite its name this olive-brown warbler-like bird only has an indistinct pale yellow wash on the belly and flanks.

**Where to see** Common resident throughout.

# Asian Glossy Starling *Aplonis panayensis* 17–20cm

This gregarious and noisy, glossy all-black starling with a conspicuous bright red eye, breeds and roosts communally in large numbers, often perching in tall dead trees and on powerlines. Immature birds look quite different, with brown upperparts and heavily streaked buffish-white and brown underparts; the iris is yellow, orange or pink, the colour becoming more intense with age. This starling is primarily frugivorous, descending on fruiting trees, sometimes in huge numbers.

**Where to see** An abundant resident in coastal areas throughout, in lowland forests, mangroves, gardens and villages, to 200m.

# Common Hill Myna
*Gracula religiosa* 30cm

These stocky, glossy black mynas, with conspicuous yellow head-wattles, are accomplished songsters with a very diverse range of songs and calls, leading to their popularity as cagebirds. In the wild the commonest call is a loud, ringing, downslurred *tee-ong*. In flight a round white wing-patch is conspicuous. The large bright yellow wattle extending from behind the eye around the nape and the orange-red bill are distinctive. A strictly arboreal frugivore and insectivore, this myna is usually seen in pairs or small flocks.

**Where to see** A common resident in lowland forests throughout, but probably declining.

# Siberian Thrush  *Geokichla sibirica*  22cm

The male of this distinctive dark slaty bluish-black thrush is distinguished by a broad, long white eyebrow and pinkish-yellow legs. The female is very different with brown upperparts, and scaly brown and buff underparts. There are records of this rare visitor to Borneo from montane forest and human habitation, but away from the island it inhabits a variety of forest types with thick understorey. It is a shy and secretive bird, usually solitary or in pairs; diet includes earthworms, insects and small fruits, for which it forages on the ground or in the understorey.

**Where to see** There are only a handful of records from Sabah and Sarawak.

# Orange-headed Thrush  *Geokichla citrina*  22cm

A shy and unobtrusive thrush that forages inconspicuously on the ground or in the understorey for earthworms, snails, insects and small fruits. The male's back is dark grey, contrasting with the striking deep chestnut-orange of the head, neck and underparts with a white bar on the wing-coverts. The female is similar, but the back is olive-grey and the wing-bar is duller.

**Where to see** A rare resident in N. Borneo, confined to a few mountains in Sabah.

# Everett's Thrush  Zoothera everetti  20cm

A very shy and secretive thrush that forages on the ground in damp mountain forests for earthworms, snails and insects. It is most often seen on quiet forest roadsides in the very early morning. This understated greyish-brown bird has a paler brown face and orange-brown underparts. Immatures are scaly brown and buff.

**Where to see** Endemic, rare resident in the mountains of W. Sabah and E. Sarawak.

# Fruit-hunter  *Chlamydochaera jefferyi*  22cm

An unobtrusive arboreal frugivore that forages in the midstorey of lower and upper montane forests. This bird is nomadic in response to fruiting events; usually found in pairs, but it may congregate in large numbers at fruiting trees. They are notably sexually dimorphic, the male being grey with a buffish-brown forehead and face, black primaries and breast-shield, and a long broad black eye-stripe. The grey is replaced by warm brown on the female.

**Where to see** An endemic, local and uncommon resident in the north-central, western and south-western mountain ranges.

♀

♀  ♂

# Oriental Magpie-Robin   *Copsychus solaris*   20cm

A familiar and much-loved garden bird on Borneo, this powerful songster can often be seen feeding on invertebrates, and occasionally small vertebrates, which it drops onto from a prominent perch. Usually seen in pairs and can be aggressive and very territorial. In the south and west the male is glossy black with a broad white stripe on the centre of the wing and white underparts. In the north and east, the underparts are all black. In both, the females are more greyish-black.

**Where to see** A common resident in coastal woodlands, mangroves, cultivated areas, villages, gardens and plantations throughout.

♀

♂

# White-crowned Shama  *Copsychus malabaricus*  22–27cm

A remarkable bird, not least for its extraordinarily rich, melodious song with a bubbly quality that is highly varied, often incorporating mimicry. The male is glossy black with a prominent white stripe over the crown, white lower back, long black tail with white outer feathers and rufous breast to vent; the female is slightly duller with a shorter tail. Despite its striking appearance it can be difficult to see as it often stays in thick cover, hopping, running and making short flights in the understorey. It is usually solitary, and cocks and fans its tail when giving its alarm call, which sounds like a breaking stick.

**Where to see** Endemic, a common resident in lowland to lower montane forest in Sabah and northern E. Kalimantan.

♂

# Grey-streaked Flycatcher
*Muscicapa griseisticta*  13cm

A migratory, solitary arboreal insectivore that sallies from high, prominent perches in lowland forest and coastal woodlands. The combination of pale lores, heavy streaks on the underparts and long wings are diagnostic.

**Where to see** A rare non-breeding visitor from September to mid-March.

# Asian Brown Flycatcher
*Muscicapa dauurica*  13cm

A greyish-brown flycatcher with white lores, a narrow pale eye-ring and whitish breast washed dirty greyish-brown. The black bill is yellowish at the base of the lower mandible. These migratory arboreal insectivores are recorded in coastal woodlands, scrub, plantations and gardens from early August to late April. They are usually solitary, foraging in the midstorey from prominent perches, sallying to catch insects on the wing.

**Where to see** An uncommon non-breeding visitor and passage migrant throughout.

# Grey-chested Jungle-Flycatcher  *Cyornis umbratilis*  15cm

A lowland specialist that sallies, hawks and gleans insects in the lower to midstorey; usually solitary. This inconspicuous flycatcher is dark brown, pale around the face with a bright white throat and pale grey underparts divided by a darker grey breast-band. The song is a repetitive, pretty *wee-tee-tee*.

**Where to see** A common resident in lowland and coastal areas throughout.

# Pale Blue Flycatcher
*Cyornis unicolor* 17cm

Only the male is pale blue, with a paler supercilium and blackish lores. This canopy-dweller of primary forests hawks insects from prominent perches and is possibly much overlooked. The female is brown with rufous-brown edges on the wings and tail, while the juvenile has buffish spots and speckles. It can be easily mistaken for Verditer Flycatcher.

**Where to see** An uncommon to rare resident in lowland forests in northern Borneo.

# Malaysian Blue Flycatcher  *Cyornis turcosus*  14cm

Many of the blue flycatchers are quite similar in appearance and care is needed when identifying them. The male of this species is cobalt-blue with a shining blue forehead and rump, blackish lores, rufous-orange breast and flanks fading to a whitish belly and undertail-coverts. The female is paler. The similar Sunda Blue Flycatcher has less extensive pale blue on the forehead, a black chin and duller blue wings. Usually found near rivers and watercourses where it forages from exposed low perches for flying insects; solitary or in pairs.

**Where to see** A common resident throughout.

# Bornean Blue Flycatcher  *Cyornis superbus*  15cm

This blue flycatcher favours deep forest, often near waterways, where it is usually seen singly or in pairs. The male is deep blue with a bright shining blue forehead and supercilium, black face and bright shining blue rump. It has more extensive bright shining blue on the head and back than any other blue-flycatcher. The female is brown with a bright rufous rump.

**Where to see** An endemic, uncommon resident in lowland forests throughout.

# Verditer Flycatcher *Eumyias thalassinus* 16cm

The male is turquoise-blue with prominent black lores and a narrow frontal band. The female is paler, with greyish-white lores. Pale Blue Flycatcher is bluer with a shorter tail, longer bill, more contrast between the blue breast and greyish-blue belly, and no white tips on the undertail-coverts. Verditer Flycatcher sallies for flying insects from exposed perches in the mid- to upper storey, often returning to the same branch.

**Where to see** An uncommon resident in lowland forest throughout.

♀

♂

# Indigo Flycatcher   *Eumyias indigo*   15cm

A pretty indigo-blue flycatcher with a bright azure forehead, black lores, greyish-white belly and yellowish undertail-coverts. The sexes are alike. They feed on insects and some berries generally in the midstorey and sometimes join mixed feeding flocks.

**Where to see** A common resident in the montane forests of the north-central ranges.

# Eyebrowed Jungle-Flycatcher *Vauriella gularis* 15cm

An unassuming flycatcher of montane forest that is usually solitary and seen foraging for insects close to the ground. Although confiding it can be easily overlooked as it perches quietly in dark forest cover. The upperparts are rich reddish-brown with white lores and a short grey eyebrow; the white throat is sharply demarcated from the grey underparts.

**Where to see** Endemic, common resident in the north-central mountain ranges.

# Blue-and-white Flycatcher
*Cyanoptila cyanomelana* 16–17cm

The male is rich cobalt-blue with a black face and upper breast and flanks, and a white belly. The female is plainer greyish-brown with rufous-brown wings. The similar White-tailed Flycatcher is not as deep blue and has a blue face. Both species are found in lowland to lower montane forests. They feed on insects and berries, typically sallying out from perches in the midstorey to catch insects on the wing.

**Where to see** An uncommon non-breeding visitor and passage migrant in N. Borneo from late October to late April.

# White-crowned Forktail *Enicurus leschenaulti* 25–28cm

A shy, restless and easily disturbed bird of fast-flowing rocky streams in primary lowland forests. It is instantly recognisable, with a bright white crest, black back, white belly and long, deeply forked tail, which is barred black and white. It actively forages for small invertebrates on rocks and banks of well-vegetated streams and occasionally on narrow roads, especially after rain. If disturbed it flies rapidly out of view while uttering its loud *screeee* call.

**Where to see** A common lowland resident but very wary and intolerant of disturbed forest.

# Bornean Whistling-Thrush
*Myophonus borneensis* 26cm

Although it appears all black in poor light, the plumage of the male is actually a glossy bluish-black with bright blue spangles and a small bright blue patch on the shoulder. The female is all blackish-brown. This inhabitant of montane forests is usually found near rocky streams, where it is solitary, foraging around rocks and in leaf-litter for invertebrates and frogs. It frequently fans its tail downwards in a unique manner. The harsh, high-pitched *cheet* can be heard clearly over the sound of fast-running water.

**Where to see** An endemic, uncommon resident in the north-central and western mountain ranges.

**Flycatchers**

# Narcissus Flycatcher
*Ficedula narcissina* 13cm

The male is a striking black and yellow with a long, broad yellowish-orange supercilium, a white patch on the wing and a bright orange-yellow rump. The female is olive-green with darker wings and greyish-white underparts. Usually seen singly or in pairs; sallies from perches in the midstorey to catch insects on the wing.

**Where to see** An uncommon non-breeding visitor and passage migrant in lowland to lower montane forests in Sabah and Sarawak.

♂

♂

# Snowy-browed Flycatcher
*Ficedula hyperythra* 11cm

A tiny slaty-blue flycatcher with a short bright white eyebrow and rufous-orange underparts. The female is greyish-brown with a pale orange forehead and eye-ring, and buffish-orange throat to belly, while the juvenile is brown with a speckled appearance. It can be tame and often inquisitive. Perches on the sides of trunks and exposed roosts to forage on or near the ground for insects.

**Where to see** A common montane resident.

# Little Pied Flycatcher
*Ficedula westermanni*   10cm

The male is a tiny black-and-white flycatcher, while the female is greyish with a brownish rump. They can be seen sallying from perches in the mid- to upper storey of the forest to catch insects on the wing and glean from foliage. Although usually solitary, they are often in pairs in the breeding season.

**Where to see** A common resident in the north-central and western mountain ranges.

♀

♂

# Rufous-chested Flycatcher
*Ficedula dumetoria*  12cm

A shy and inconspicuous flycatcher of the lowlands, usually seen singly or in pairs, foraging for insects in the dense lower to midstorey by sallying from a perch. Sometimes it also gleans from foliage in a warbler-like fashion. The male is slaty-black with a long white supercilium and a prominent white stripe on the wings, and a burnt-orange breast. The female is dull olive-brown with the breast, rump and tail washed rufous.

**Where to see** A locally common resident in primary lowland to hill forests throughout.

# Lesser Green Leafbird
*Chloropsis cyanopogon*  15cm

Although very similar to Greater Green Leafbird, the male has a narrow yellow border around the black face and throat. The female is all grass-green with an inconspicuous blue malar stripe. They are generalist feeders, foraging in the mid- to upper storey and often in mixed feeding flocks.

**Where to see** A common resident in lowland forests throughout.

# Greater Green Leafbird
*Chloropsis sonnerati*   20cm

A bright green forest bird. The male
has a black mask-like face and throat
with a glossy dark blue malar stripe and
a turquoise-blue shoulder patch. The
female is like the male but lacks the mask
and has a bright yellow eye-ring and a
yellow throat with only a faint blue malar
stripe. The very similar Lesser Green
Leafbird is smaller, and the male has a
narrow, indistinct yellow border around
the black face and throat, while the
female lacks yellow on the throat.

**Where to see** A common resident of
lowland forests throughout.

# Bornean Leafbird
*Chloropsis kinabaluensis*   18cm

This endemic leafbird is found only in the
mountains of the north and is the only
leafbird in which the female has a black
mask. The male's black mask is bordered
yellowish-green, while the female's is
bordered bluish-green and lacks the
male's blue malar stripe. Both have blue-
washed wings and tail. Immature birds
lack the black mask. This species may be
confused with male Blue-winged Leafbird
but it is not found in the lowlands or
in the south (except for the Meratus
Mountains).

**Where to see** An endemic, common
montane resident in the north-central
ranges and the Meratus Mountains.

**Fairy-bluebirds**

# Asian Fairy-bluebird  *Irena puella*  21–26cm

An unmistakable, startlingly deep blue and velvet-black bird with red eyes. The female is a duller turquoise-blue with blackish wings and tail, while the immature is like the female but duller with brownish-tinged wings. Although primarily frugivorous it will sometimes take insects, often joining mixed feeding flocks. The memorable song is a rapid, loud, rollicking *hwit-hwit-hwit-hwit*.

**Where to see** A common resident throughout in primary and secondary lowland forests.

♀

♂

# Yellow-rumped Flowerpecker *Prionochilus xanthopygius* 9cm

This endemic blue-and-yellow flowerpecker is common in lowland forest where it forages for small fruits, nectar and tiny invertebrates at all levels. The brighter male is very like Crimson-breasted Flowerpecker, which lacks the distinctive yellow rump, has a more diffuse red breast-patch and a white moustachial stripe.

**Where to see** A common endemic resident in N. Borneo.

**Flowerpeckers**

# Yellow-breasted Flowerpecker  *Prionochilus maculatus*  10cm

A tiny but chunky dark olive-green bird with a stout bill and a conspicuous orange-red patch on the crown, boldly streaked olive-green and yellow underparts, and a dark red eye. It is usually solitary or in pairs and forages for small berries and nectar at all levels in the forest.

**Where to see** A common resident in lowland forest and gardens throughout.

# Yellow-vented Flowerpecker  *Dicaeum chrysorrheum*  10cm

This uncommon resident is olive-green with blackish wings and tail; the white underparts are boldly streaked blackish-olive and the undertail-coverts are bright yellow. It is found in lowland and hill forests where it feeds on small fruits, especially mistletoe, nectar and small invertebrates at all levels.

**Where to see** Uncommon resident throughout.

## Orange-bellied Flowerpecker *Dicaeum trigonostigma* 8cm

A familiar garden bird on Borneo, the male has mostly dark blue upperparts with an orange back and a bright orange breast and belly. The female is dark olive-brown with a dull orange rump. These flowerpeckers can be seen in lowland to lower montane forests, as well as plantations and gardens. They often feed on tiny fruits, especially mistletoe, and small invertebrates, with nectar constituting an important part of their diet.

**Where to see** A common resident throughout.

## Scarlet-backed Flowerpecker *Dicaeum cruentatum* 8cm

A boldly coloured black-and-red flowerpecker that favours secondary scrub, mangroves, plantations and gardens in the lowlands. The male is black with a bright red crown, back and rump. The female is olive-brown with a less conspicuous bright red lower back and rump. They are usually in pairs or small groups and feed on small fruits, especially mistletoe, seeds, nectar and small invertebrates at all levels.

**Where to see** A common resident throughout.

# Black-sided Flowerpecker  *Dicaeum monticola*  8cm

Often seen as black dots hurtling high overhead, uttering a high-pitched, metallic *zit*, these hill and mountain specialists are endemic to Borneo. The male is dark glossy blackish-blue with a bright crimson throat and upper breast bordered blackish, while the female is olive-green with a yellowish wash on the rump. They feed on small fruits, seeds, nectar and invertebrates.

**Where to see** An endemic, common resident in north-central mountain ranges.

# Ruby-cheeked Sunbird  *Chalcoparia singalensis*  10cm

This small, brightly coloured bird is named for the male's copper-red cheeks. It also has a bright metallic green back and apricot-orange breast, while the female is olive-brown, with a more washed-out orange breast. For a sunbird it has a relatively short, straight bill. It feeds actively on small invertebrates, fruits and nectar, and is often seen in gardens as well as lowland to hill forests, mangroves and plantations.

**Where to see** Common resident throughout.

# Brown-throated Sunbird *Anthreptes malacensis* 13cm

A relatively large sunbird, the male has a metallic blue-green back, brown cheeks and yellow underparts. In bright light it appears very colourful and is named for the male's brownish-purple throat. The female is much plainer with olive-green upperparts and pale yellow underparts. It feeds mainly on invertebrates, but also nectar and over-ripe fruits, favouring open and disturbed habitats.

**Where to see** Common resident throughout in all types of forested areas, including gardens.

# Red-throated Sunbird *Anthreptes rhodolaemus* 12cm

Very similar to Brown-throated Sunbird but the male has a more reddish throat, cheeks and wings. The female is also similar but has a narrower yellow eye-ring and less yellow underparts. This species gleans leaves for small invertebrates at all levels and also feeds on nectar and over-ripe fruits. It can occur in all types of forested habitats but possibly favours more intact inland wooded habitats in the lowlands.

**Where to see** A scarce resident throughout.

# Olive-backed Sunbird *Cinnyris jugularis* 11cm

This very familiar garden bird can be found in all types of lowland forested habitats including mangroves and plantations. The male, with his olive-green back, yellow belly and blackish-purple metallic throat is easily identified. The female is like the male but lacks the metallic blackish-purple in her plumage. They feed primarily on nectar as well as small invertebrates and are quite aggressive.

**Where to see** Common resident throughout.

# Copper-throated Sunbird *Leptocoma clacostetha* 13cm

This very distinctive sunbird could not be mistaken for any other on Borneo. The male's plumage is metallic green on the upperparts with glossy black wings and tail, metallic coppery-red throat and metallic purplish-blue breast. The female is plain olive-green with a broken white eye-ring and yellow underparts. They are commonest in mangroves where they feed at mangrove flowers and on small invertebrates; actively flit from flower to flower at all levels.

**Where to see** Locally common resident in coastal areas throughout.

# Temminck's Sunbird *Aethopyga temminckii* 13cm (male); 10cm (female).

Could be mistaken for Crimson Sunbird but the male has a long scarlet tail and yellow rump, while the female has olive-green upperparts and a reddish wash on the wings and tail. A very active bird, mostly in hill and montane forest but sometimes lower. Feeds primarily on small invertebrates but also nectar and little fruits.

**Where to see** A common resident, patchily distributed throughout.

# Crimson Sunbird *Aethopyga siparaja* 12–15cm

A bright crimson-red lowland sunbird with a glossy purple forehead, olive-grey wings and belly, and a long black tail. The female is plain greyish-olive with a yellow wash on the belly. Feeds on insects and nectar, usually in the lower storey and often in pairs.

**Where to see** Common resident throughout in lowland forests, coastal scrub, mangroves, plantations, and gardens.

♂

# Purple-naped Spiderhunter *Kurochkinegramma hypogrammicum* 14cm

A mostly olive-green bird, the male has an inconspicuous metallic purple-blue nape, and lower back and rump (often invisible in the field), and the underparts are yellower with broad olive streaks. The female lacks the metallic purple-blue on the nape and rump. They are found in well-forested areas where they forage for small invertebrates, fruit and nectar in the lower to midstorey, especially favouring banana flowers.

**Where to see** Locally common but inconspicuous resident throughout.

♂

# Thick-billed Spiderhunter *Arachnothera crassirostris* 17cm

A dark olive-green forest bird with a broken yellow eye-ring and yellow underparts. The remarkably long, decurved bill is used for probing flowers and mistletoe for invertebrates and nectar, usually in the midstorey. The bill is thicker than that of other spiderhunters and this species lacks the streaky belly of the similar Bornean and Grey-breasted Spiderhunters.

**Where to see** Uncommon lowland to lower montane resident throughout.

# Spectacled Spiderhunter *Arachnothera flavigasta* 22cm

A large spiderhunter with a triangular yellow ear-patch, broad yellow eye-ring and pinkish legs. The very similar Yellow-eared Spiderhunter is smaller with a longer, finer bill, larger tufted ear-patch and narrower eye-ring. It favours disturbed areas in primary and secondary lowland forests and gardens where it feeds on invertebrates, fruit and nectar.

**Where to see** An uncommon to rare resident, patchily distributed throughout.

# Long-billed Spiderhunter  *Arachnothera robusta*  22cm

A very aggressive spiderhunter that feeds predominantly on invertebrates and nectar in the upper storey of primary and secondary forests. Fittingly, its bill is the longest of any of the spiderhunters. This bird is olive-green with a streaky yellowish-green throat and breast. The similar Little, Thick-billed, Spectacled and Yellow-eared Spiderhunters all lack streaking on the underparts, and Bornean and Grey-breasted Spiderhunters lack yellow on the underparts.

**Where to see** Uncommon lowland to lower montane resident, patchily distributed throughout.

# Little Spiderhunter  *Arachnothera longirostra*  15cm

The smallest and most commonly encountered spiderhunter, it is also the only one with a whitish-grey throat. It is very active, flying rapidly through the lower storey where it feeds on nectar and small invertebrates, often frequenting spiders' webs. It pierces the base of flowers, especially those of banana and ginger, to extract nectar.

**Where to see** An abundant resident throughout in lowland and hill forests, mangroves, plantations and gardens.

**Sunbirds and Spiderhunters**

# Yellow-eared Spiderhunter
*Arachnothera chrysogenys*  18cm

A medium-sized spiderhunter with a
long fine bill and a large yellow cheek
patch of elongated feathers. It is similar
to Spectacled Spiderhunter but can be
told from all other spiderhunters by the
streaking on the underparts and the
narrow yellow eye-ring. It consumes
insects, fruit and nectar, foraging high in
the canopy where it hovers, flutters and
hangs upside-down.

**Where to see** An uncommon resident
throughout in lowland and hill forests,
mangroves, plantations and gardens.

# Bornean Spiderhunter
*Arachnothera everetti*  21cm

A chunky dark olive-green spiderhunter
with streaky grey-and-cream underparts
and flesh-pink legs. The similar Grey-
breasted Spiderhunter is smaller with
a stouter, shorter bill and a greyer, less
streaky belly. It feeds on nectar and small
invertebrates and is highly mobile, flying
rapidly through the lower storey in search
of flowers.

**Where to see** Endemic, uncommon
lower montane resident in N. and C.
Borneo.

# Whitehead's Spiderhunter

*Arachnothera juliae* 17cm

A large, distinctive, dark brown spiderhunter with a heavily whitish-streaked head and underparts, and a conspicuous bright yellow vent and uppertail-coverts. It is a montane resident, usually seen singly or in pairs in the canopy, but sometimes in the mid-canopy; relatively sedate. Its very distinctive call is a wheezy *wee-chit* and it also utters a complex series of nasal but wheezy twitters and trills.

**Where to see** Endemic, locally common resident in the north-central mountain ranges.

## Grey-breasted Spiderhunter *Arachnothera modesta* 18cm

Although very similar to Bornean Spiderhunter, there is little overlap in their ranges. This uncommon resident is patchily distributed in Sarawak, Brunei and Kalimantan in lowland forest, while the previous species is mostly found in Sabah in hill and montane forests. It feeds actively on invertebrates and nectar and flies low and fast through the understorey and midstorey.

**Where to see** Uncommon in lowland forests and gardens everywhere except N. Borneo.

## Eurasian Tree Sparrow *Passer montanus* 15cm

A very familiar bird in many parts of the world, this sparrow has a chestnut-brown hood, white cheeks with a black spot, and streaked black-and-rufous upperparts. Pinkish legs and feet.

**Where to see** An abundant introduced resident around human habitation throughout.

# Dusky Munia *Lonchura fuscans* 11cm

A small, all-dark brown finch-like bird with a contrasting triangular, pale grey bill. The legs and feet are bluish-grey. It can be seen feeding on rice and grass seeds at forest edges and in secondary forest, paddyfields, grasslands, young plantations and gardens. It spends most of its time on the ground and low in vegetation.

**Where to see** Essentially an endemic (there is a small population on a Philippines island close to the Bornean coast), a common to abundant resident throughout.

# Scaly-breasted Munia *Lonchura punctulata* 10cm

A small, dark ginger-brown munia with a distinctive scaly breast. It frequents paddyfields and long grass at sea level. Usually in small to large flocks, feeding on seeds in low grass.

**Where to see** A recent colonist, locally common in the north but patchy elsewhere.

# Black-headed Munia *Lonchura atricapilla* 12cm

A medium-sized chestnut munia with a distinctive black hood and lead-grey bill. These munias can be seen in long grass, scrub, paddyfields and overgrown plantations. Highly sociable, they are usually encountered in large to very large flocks.

**Where to see** An abundant resident throughout.

# Grey Wagtail *Motacilla cinerea* 18cm

This long-tailed grey-and-yellow bird with pink legs can be seen on narrow wooded rocky streams and roads where it actively forages on the ground, darting and running for small invertebrates. It constantly pumps its tail up and down when standing still.

**Where to see** A common migrant throughout, from August to early May.

# Paddyfield Pipit *Anthus rufulus* 16cm

A bird of grasslands, fields, paddyfields, plantations and airfields, the plumage is dark brown fringed buffish-white with warm brown wings and a rufous wash on the belly and flanks. It forages for small invertebrates on open ground at sea level, usually singly or in pairs. When it takes flight the white outer tail feathers are conspicuous.

**Where to see:** A common (Sabah) to uncommon (Brunei, Sarawak, E. Kalimantan) resident in N. Borneo.

# FURTHER READING AND RESOURCES

## Societies and Clubs

### Malaysian Nature Society (MNS)
Established in 1940, the MNS is one of the oldest environmental organisations in Malaysia. It has branches in most of the states in Malaysia, including on Borneo where there are two in Sabah and two in Sarawak. The Society is active in all areas of conservation and scientific research in Malaysia and has conducted many successful environmental activities and campaigns. It is a voluntary, membership-based organisation with approximately 3,800 members.

### WWF-Malaysia
The Malaysian branch of the World Wide Fund for Nature actively works on conservation issues throughout Malaysia, including in Sabah and Sarawak. Its stated goals include efforts to conserve Malaysia's forests, oceans, wildlife, food, climate and energy, and freshwater environments with particular attention to the three key drivers of environmental problems: markets, finance and governance.

### Burung Indonesia
The Indonesian branch of Birdlife International aims to be the guardian of Indonesia's wild birds and their habitats through working with people for sustainable development. The organisation's stated goals are to promote the conservation of sites, species and habitats; to work with communities to promote collaborative conservation and natural resource management for sustainable development; and to improve management of habitats, sites, and species.

### Oriental Bird Club (OBC)
Established in 1985, the OBC has a worldwide membership and publishes the scientific *Journal of Asian Ornithology* as well as the *BirdingASIA* bulletin. The aims of the club are to encourage an interest in wild birds of the Oriental region and their conservation; to promote the work of regional bird and nature societies; and to collate and publish information on Oriental birds

### Borneo400
The Borneo400 is a club for people who have seen at least 400 bird species in Borneo. Contact ckleong@borneobirds.com.

## Books

Francis, C.M. 1984. *Pocket Guide to the Birds of Borneo*. The Sabah Society, Malaysia.

Lai, F. and Olesen, B. 2016. *A Visual Celebration of Borneo's Wildlife*. Periplus Editions, North Clarendon, VT.

Myers, S. 2016. *Birds of Borneo* (2nd edition, Helm Field Guides). Helm, London.

Smythies, B.E. 1999. *The Birds of Borneo*, edited by G.W.H. Davison. Natural History Publications, Borneo.

Yong, D.L. 2018. *The 125 Best Birdwatching Sites in Southeast Asia*. John Beaufoy Publishing, Oxford.

# Blogs

'Birders of Borneo Island' by Jason Bugay Reyes
https://horukuru.blogspot.com

'Borneo Birds' by CK Leong
https://borneobirds.com

'Birding and Photography' by Denis Degullacion
http://degullacion.blogspot.com (last updated May 2015)

'Amazing Borneo' by Jordan Sitorius
http://amazingborneo.blogspot.com

# Websites

Burung Nusantara – birdwatching and bird conservation in Indonesia.
http://burung-nusantara.org

Borneo Bird Images
http://www.borneobirdimages.com

Brunei Nature Society – an NGO that aims to promote an interest in, and to study, natural history in general and that of Brunei Darussalam in particular.
http://www.nzboyce.org/brunatsoc/bruneinaturesociety.org

eBird – any birding sites in Borneo are covered here, with species lists and locations.
ebird.org/region

*International Ornithological Congress* (IOC) World Bird List
worldbirdnames.org

# Photo credits

Bloomsbury Publishing would like to thank the following for providing photographs and permission to use copyright material (ordered left–right, top–bottom).

SM = Susan Myers; LWK = Liew Weng Keong; CB = Carlos Bocos; JE = James Eaton; RH = Rob Hutchinson; DB = David Bakewell; JJH = John and Jemi Holmes.

**Front cover** all SM; **back cover** LWK, LWK, LWK, SM; **1** JE; **3** LWK; **4** SM; **7** LWK, LWK; **9** all SM; **10** SM; **11** SM; **12** SM; **13** LWK, LWK; **14** SM, SM; **15** SM; **16** SM; **17** SM; **18** SM; **19** LWK; **20** SM, LWK; **23** SM, CB; **24** LWK, LWK; **25** LWK, LWK; **26** CB, LWK, LWK; **27** LWK, LWK; **28** SM, LWK; **29** CB, SM; **30** LWK, LWK; **31** CB, LWK; **32** LWK, SM; **33** LWK, LWK; **34** LWK, LWK; **35** SM, LWK; **36** SM, LWK, SM; **37** LWK, SM; **38** LWK, LWK; **39** LWK, JE; **40** all LWK; **41** LWK, LWK; **42** LWK, LWK; **43** JE, JE; **44** LWK, LWK; **45** RH, LWK; **46** LWK, SM; **47** LWK, LWK; **48** LWK, LWK; **49** SM, CB; **50** CB, SM; **51** CB, CB; **52** CB, SM; **53** LWK, LWK; **54** all LWK; **55** CB, CB; **56** CB, CB; **57** CB, CB; **58** CB, CB; **59** CB, SM, SM; **60** SM, SM; **61** CB, SM; **62** CB, CB; **63** SM, CB; **64** CB, CB; **65** SM, SM; **66** CB, LWK; **67** LWK, LWK; **68** LWK, LWK; **69** LWK, LWK; **70** SM, CB; **71** CB, SM; **72** LWK, LWK; **73** LWK, LWK; **74** SM, SM; **75** LWK, LWK; **76** LWK, LWK; **77** CB, LWK; **78** SM, CB; **79** CB, CB; **80** SM, SM; **81** JE, LWK; **82** all LWK; **83** all LWK; **84** LWK, CB; **85** SM, LWK, SM; **86** LWK, LWK; **87** LWK, LWK; **88** LWK, LWK; **89** LWK, SM; **90** LWK, JE ; **91** SM, SM; **92** all LWK; **93** LWK, LWK; **94** LWK, LWK; **95** SM, SM; **96** LWK, LWK; **97** LWK, LWK; **98** CB, LWK; **99** LWK, LWK; **100** LWK, LWK; **101** LWK, LWK; **102** LWK, LWK; **103** LWK; **104** SM, LWK; **105** SM, LWK; **106** LWK, LWK; **107** SM, LWK; **108** SM, SM; **109** LWK, LWK; **110** LWK, LWK; **111** LWK, LWK; **112** LWK; **113** LWK, LWK; **114** CB, LWK; **115** RH; **116** LWK, LWK; **117** SM, LWK; **118** LWK, LWK; **119** SM, SM; **120** SM, LWK, LWK; **121** SM, LWK; **122** LWK, LWK; **123** LWK, LWK; **124** CB, LWK; **125** LWK, LWK; **126** LWK, RH; **127** LWK, LWK, SM; **128** JE, LWK; **129** LWK, LWK; **130** SM, LWK; **131** LWK, LWK; **132** LWK, CB; **133** RH, SM; **134** LWK, SM; **135** all LWK; **136** LWK, LWK; **137** LWK, LWK; **138** LWK, SM, LWK; **139** SM, LWK; **140** CB, LWK; **141** LWK, LWK; **142** LWK, SM; **143** SM, LWK; **144** JE, LWK; **145** LWK, JE; **146** LWK, LWK; **147** SM, LWK; **148** LWK, CB; **149** LWK, LWK; **150** LWK, LWK; **151** LWK, CB; **152** RH, CB; **153** SM, RH; **154** CB, SM; **155** LWK, LWK; **156** CB, LWK; **157** LWK, JE; **158** LWK, LWK; **159** LWK, SM; **160** SM, LWK; **161** CB, LWK; **162** LWK, CB; **163** SM, LWK; **164** LWK, LWK; **165** LWK, LWK; **166** LWK, LWK; **167** SM, LWK; **168** SM, LWK; **169** LWK, SM; **170** SM, JE; **171** JE, LWK, JE; **172** LWK, LWK; **173** LWK, LWK; **174** SM, LWK; **175** LWK, LWK; **176** CB, CB; **177** LWK, JE; **178** LWK, CB; **179** LWK, SM; **180** SM, SM; **181** LWK, LWK; **182** LWK, SM; **183** JJH, LWK, CB; **184** LWK, LWK; **185** LWK, LWK; **186** LWK, JE; **187** LWK, LWK; **188** SM, CB; **189** CB, CB; **190** LWK, CB; **191** LWK, SM; **192** LWK, LWK; **193** LWK, SM; **194** CB, SM; **195** LWK, SM; **196** SM, LWK; **197** LWK, RH; **198** CB, CB; **199** LWK, JE; **200** LWK, LWK; **201** LWK, SM, LWK, LWK; **202** LWK, LWK; **203** LWK, JE; **204** LWK, LWK; **205** SM, LWK; **206** LWK, LWK; **207** SM, LWK; **208** LWK, LWK; **209** LWK, RH; **210** DB, CB; **211** LWK, LWK; **212** LWK, LWK; **213** LWK, LWK; **214** CB, JE; **215** LWK, LWK; **216** LWK, CB; **217** LWK, CB; **218** CB; **219** CB, CB.

# INDEX

# Orange-bellied Flowerpecker *Dicaeum trigonostigma* 8cm

A familiar garden bird on Borneo, the male has mostly dark blue upperparts with an orange back and a bright orange breast and belly. The female is dark olive-brown with a dull orange rump. These flowerpeckers can be seen in lowland to lower montane forests, as well as plantations and gardens. They often feed on tiny fruits, especially mistletoe, and small invertebrates, with nectar constituting an important part of their diet.

**Where to see** A common resident throughout.

# Scarlet-backed Flowerpecker *Dicaeum cruentatum* 8cm

A boldly coloured black-and-red flowerpecker that favours secondary scrub, mangroves, plantations and gardens in the lowlands. The male is black with a bright red crown, back and rump. The female is olive-brown with a less conspicuous bright red lower back and rump. They are usually in pairs or small groups and feed on small fruits, especially mistletoe, seeds, nectar and small invertebrates at all levels.

**Where to see** A common resident throughout.

# Black-sided Flowerpecker *Dicaeum monticola* 8cm

Often seen as black dots hurtling high overhead, uttering a high-pitched, metallic *zit*, these hill and mountain specialists are endemic to Borneo. The male is dark glossy blackish-blue with a bright crimson throat and upper breast bordered blackish, while the female is olive-green with a yellowish wash on the rump. They feed on small fruits, seeds, nectar and invertebrates.

**Where to see** An endemic, common resident in north-central mountain ranges.